Polyamide Thin Film Composite Membranes for Water Applications

The global market for polymeric membranes used in water and wastewater treatment is experiencing robust growth, with polyamide (PA) thin film composite (TFC) membranes dominating reverse osmosis (RO) and nanofiltration (NF) processes. This monograph presents the latest trends in characterization techniques for PA TFC membranes and provides the most current and relevant information on these techniques tailored specifically for TFC NF and RO membranes.

Features

- Focuses solely on the characterization of PA TFC membranes used in water and wastewater treatment.
- Provides the latest insights into employing advanced and emerging characterization tools to analyze the intrinsic properties of the PA selective layer and evaluate the overall performance of PA TFC membranes during filtration processes.
- Extensively examines the strengths and limitations of each membrane characterization tool, offering in-depth analyses for readers.

This book is an indispensable reference and practical guide for advanced students, researchers, and scientists involved in NF and RO membrane fabrication and characterization, including those in the fields of chemical, materials, and environmental engineering.

Polyamide Thin Film Composite Membranes for Water Applications

Advanced Characterization Techniques

Ying Siew Khoo, Jia Zheng Oor,

Woei Jye Lau, and Jia Wei Chew

CRC Press

Taylor & Francis Group

Boca Raton London New York

CRC Press is an imprint of the
Taylor & Francis Group, an **informa** business

First edition published 2024
by CRC Press
2385 NW Executive Center Drive, Suite 320, Boca Raton FL 33431

and by CRC Press
4 Park Square, Milton Park, Abingdon, Oxon, OX14 4RN

CRC Press is an imprint of Taylor & Francis Group, LLC

© 2024 Ying Siew Khoo, Jia Zheng Oor, Woei Jye Lau, Jia Wei Chew

ISBN: 978-1-032-68856-5 (hbk)
ISBN: 978-1-032-69033-9 (pbk)
ISBN: 978-1-032-69034-6 (ebk)

DOI: 10.1201/9781032690346

Typeset in Times
by SPi Technologies India Pvt Ltd (Straive)

Contents

Preface

The book aims to present a comprehensive and up-to-date overview of advanced and emerging characterization techniques for thin film composite (TFC) membranes. Specifically, it highlights the potential of using these advanced techniques to analyze the intrinsic properties of the polyamide (PA) selective layer of TFC membranes. Comprehending the attributes of the PA layer is essential when aiming to improve membrane performance in water or wastewater treatment. This significance extends beyond pressure-driven processes like nanofiltration (NF) and reverse osmosis (RO) to include engineered osmosis processes like forward osmosis (FO) and pressure retarded osmosis (PRO). Additionally, this book emphasizes the utilization of in-situ/online characterization tools to investigate membrane fouling tendencies. By employing these tools, a deeper understanding of the mechanisms underlying fouling formation can be gained, providing valuable insights for mitigating and addressing this critical issue in membrane technology. This is the first book solely dedicated to PA TFC membrane characterization and is aimed to serve as a valuable reference and practical guide for students (especially final year undergraduates and postgraduates), membrane science and technology researchers and scientists focused on membrane fabrication/modification, new membrane materials synthesis, membrane characterization and membrane performance evaluation. By presenting these advanced techniques, the book seeks to contribute significantly to the understanding and improvement of TFC membrane technology.

About the Authors

Dr. Ying Siew Khoo received her B. Eng. (Hons) in Chemical Engineering from Universiti Malaysia Pahang (UMP) and later earned her PhD in Chemical Engineering from Universiti Teknologi Malaysia (UTM). During her doctoral studies, she worked with her supervisor – Assoc Prof Dr Woei Jye Lau – on research focused on enhancing the desalination process through surface modification of thin film nanocomposite membranes. Her contributions to the field have resulted in the publication of 17 works, including articles and book chapters, where she served as both first and co-author. Currently, Ying Siew holds the position of a postdoctoral fellow at Nanyang Technological University (NTU), Singapore.

Jia Zheng Oor is currently a final year PhD student at Nanyang Technological University (NTU), from the School of Chemistry, Chemical Engineering and Biotechnology (CCEB). She obtained her Bachelor of Engineering with Honors in Pharmaceutical Engineering from Singapore Institute of Technology (SIT) and attained her Diploma in Pharmaceutical Sciences from Temasek Polytechnic. Jia Zheng joined Professor Chew Jia Wei's membrane group for her PhD and her research mainly revolves around membrane-based pharmaceutical separations. She focuses on membrane surface modification and fabrication of active layer, with the goal of optimizing chiral purification using covalent organic framework and thin film nanocomposite membranes. To date, she has published five papers as a first and co-author. Jia Zheng has also participated in multiple conferences such as the Green & Sustainable Manufacturing (GSM) symposium and the 13th International Congress on Membranes and Membrane Processes (ICOM 2023). Her research has been recognized by the World Association of Membrane Societies (WA-MS) as she was presented an award at ICOM 2023, in Japan. In her 10 years of studying and carrying out research relevant to the pharmaceutical industry, she has remained driven and passionate about drug optimization and purification, all with the goal of improving drug chiral selection through membrane technology.

Dr. Woei Jye Lau is currently an associate professor in the Faculty of Chemical and Energy Engineering, at the Universiti Teknologi Malaysia (UTM). Dr Lau has a very strong research interest in the field of membrane science and technology for water applications and has published more than 300 scientific papers with total number of Scopus citations exceeding 13,000. He is the author of the book Nanofiltration Membranes: Synthesis, Characterization and Applications published by CRC Press and has edited several books published by CRC Press and Elsevier. Currently, Dr Lau serves as an editor for Water Reuse (International Water Association, IWA) and is subject editor for Chemical Engineering Research and Design (Elsevier). He holds a position on the Management Committee of the IWA Specialist Group on Membrane Technology. In recognition of his research excellence, Dr Lau received the Australian Endeavour Research Fellowship in 2015, UI-RESOLV Program 2016 (Indonesia), Mevlana International Exchange Program 2017/2018 (Turkey), Sakura Exchange Program 2018 (Japan), TÜBİTAK's Fellowships for Visiting Scientists 2018 (Turkey), ASEAN-India Collaborative R&D Scheme 2019, the 6th Science & Technology Exchange Program (Iran), AUN/SEED-Net's RC Grant Award 2022 and the Top Research Scientists Malaysia (TRSM) 2023.

Dr. Jia Wei Chew is a Professor of Chemical Engineering at Chalmers University of Technology in Sweden. Prior to this, she was the Technical Director at Particulate Solid Research Incorporated in Chicago. And before that, she was an Associate Professor in the School of Chemical and Biomedical Engineering at the Nanyang Technological University (NTU) in Singapore, with research focusses on fluidization, membrane-based separations, and machine learning. She has over 260 publications, ten patents/technical disclosures, and is on the editorial boards for various peer-review journals including Journal of Membrane Science Letters, Advanced Powder Technology, AIChE Journal, Membranes, Frontiers in Chemical Engineering and Scientific Reports. Her research has been recognized by various awards, including the AIChE (American Institute of Chemical Engineers) Particle Technology Forum (PTF) Best PhD Dissertation Award in 2013, the Singapore Youth Award in 2015 and the AIChE PTF Sabic Young Professional Award in 2017. She obtained her PhD degree in Chemical Engineering from the University of Colorado at Boulder, and both her bachelor's and master's degrees in Chemical Engineering from the National University of Singapore (NUS).

Introduction

1

1.1 POLYAMIDE THIN FILM COMPOSITE MEMBRANE

In the early 1970s, Cadotte and his team introduced a ground-breaking innovation in the field of desalination membranes. They developed the first polyamide (PA) thin film composite (TFC) membrane as an alternative to overcome the shortcomings of the asymmetric cellulose triacetate reverse osmosis (RO) membrane [1]. The RO membrane initially pioneered by Loeb and Sourirajan was hampered by two primary issues, namely, low water permeability and poor biological and chemical stability.

Inspired by the interfacial polymerization (IP) technique introduced by Morgan in the late 1950s [2], Cadotte and his team successfully developed an ultrathin cross-linked PA layer on the surface of a microporous substrate, resulting in the development of a high-flux TFC membrane [3]. The polymerization reaction occurs between two immiscible active monomers, namely, the amine monomer dissolved in an aqueous solution and the acyl chloride monomer dissolved in an organic solvent, right at the surface of the substrate. This chemical reaction gives rise to an ultrathin cross-linked layer, typically spanning several hundreds of nanometers. This thin layer possesses distinctive characteristics that serve to enhance water transport while simultaneously ensuring outstanding solute removal capabilities.

The use of the IP technique in TFC membrane production has become the prevalent practice in the industry. It is employed to manufacture RO membranes that exhibit exceptional efficiency in eliminating dissolved ions from seawater or brackish water, often achieving rejection rates exceeding 99%. Furthermore, these IP-based TFC membranes offer significantly higher water permeability compared to asymmetric membranes, such as those made from cellulose using

DOI: 10.1201/9781032690346-1

the traditional phase inversion method. At present, the PA TFC membranes have established a dominant presence, not just in the RO market but also in the realm of nanofiltration (NF) applications. Over the past decade, there has been a consistent and significant surge in the development and deployment of TFC membranes. These membranes have played a pivotal role in facilitating dependable desalination processes for producing clean and potable water. A report indicated that the global market for water desalination is estimated to reach \$21.7 billion by 2027, representing a significant increase from \$14.7 billion in 2022 [4]. Prominent participants in the global TFC membrane manufacturing landscape include industry leaders such as DuPont FilmTec™, Toray™ and GE Osmonics Desal. Additionally, other major players contributing significantly to the market include Synder™, TriSep™ and LG Chem's NanoH₂O™.

In the context of global research and development efforts, it is noteworthy that research publications in the fields of RO and NF have experienced a remarkable surge over the last two decades. As of October 30, 2023, data sourced from Scopus reveals that approximately 20,000 research papers pertaining to RO and NF membranes have been documented during the recent 20-year period from 2003 to 2022, with most of these studies published in international peer-reviewed journals, including *Desalination* (17.4%), *Journal of Membrane Science* (14.1%), *Desalination and Water Treatment* (5.1%) and *Separation and Purification Technology* (4.5%). Of the total number of papers published, about 40% was during the last five years (between 2018 and 2022). This substantial body of recent work underscores the growing interest and advancements in these membrane technologies.

Over the years, the performance of TFC membranes has been continuously improved not only to overcome the inevitable trade-off between selectivity and water permeability but also to improve their resistance against organic/inorganic fouling, chlorine attack and scaling. Numerous strategies have been explored to enhance the surface characteristics of the PA layer of TFC membranes. The introduction in 2007 of a new membrane type, referred to as the thin film nanocomposite (TFN) membrane, marks a significant development in membrane technology [5]. This innovation involves the integration of inorganic nanofillers into or onto the PA matrix, either during or after the IP process. TFN membranes have shown promising outcomes in mitigating the limitations associated with conventional TFC membranes. Since the performance of the PA layer is greatly influenced by the physicochemical characteristics of the active layer itself, it is of paramount importance to comprehend how modifications in the intrinsic properties can lead to performance enhancements during the filtration process.

Comprehensive characterization at the nanometer scale and even at the molecular level serves as a crucial tool in gaining invaluable insights into membrane properties and behavior. These insights play a pivotal role in the advancement of water desalination and wastewater treatment technologies. Advanced characterization techniques empower researchers to delve deeper into critical factors such as membrane structure and morphology, surface chemistry and interactions with a diverse range of solutes. The information gained from advanced characterization has the potential to catalyze the development of an advanced TFC membrane that can effectively address not only the permeability-selectivity trade-off issue but also tackle fouling problems encountered during membrane operation. This underscores the essential role that cutting-edge characterization methods play in pushing the boundaries of membrane technology and the corresponding applications in water treatment and desalination.

Common Membrane Characterization Techniques

2

2.1 OVERVIEW OF COMMON MEMBRANE CHARACTERIZATION TECHNIQUES

Given the inherent diversity among PA TFC membranes with regard to materials (comprising various monomer types and additives), morphology (which may include ridge-and-valley or leaf-like structures), transport/separation mechanisms (encompassing Donnan exclusion, sieving and/or solution-diffusion mechanisms), configurations (ranging from flat sheets to hollow fibers) and applications (including both pressure-driven and osmotically driven processes), a wide spectrum of techniques and methods becomes essential for their thorough characterization.

The characterization methods for PA TFC membranes are typically classified into three main categories: (i) surface chemistry, including PA cross-linking degree and surface charge properties; (ii) physical properties, such as surface morphology, pore size and roughness; and (iii) membrane separation performance, including water permeability and solute rejection. Depending on the intended application of PA TFC membranes, additional stability tests may be conducted to assess their durability in specific environments, such as exposure to chlorination, thermal conditions and compaction.

DOI: 10.1201/9781032690346-2

When it comes to commercial PA TFC membranes available in the market, membrane manufacturers generally offer limited information to the public. Typically, only data such as permeate flow rate, salt rejection under specific testing conditions, tolerance to free chlorine, pH range tolerance and maximum operating temperature and pressure are provided. However, it must be noted that this information may not provide a comprehensive understanding of the performance of PA TFC membrane and its surface interaction with feed solutes when dealing with feed characteristics beyond those typically encountered in brackish and seawater desalination processes.

Table 2.1 provides an overview of the analytical instruments and methods commonly employed in the laboratory for characterizing PA TFC membranes,

TABLE 2.1 Analytical instruments/methods used to access the surface characteristics and performance of PA TFC membranes for water application

PROPERTIES ASSESSMENT	TYPE OF INSTRUMENT/ METHOD	FUNCTION(S) OF INSTRUMENT/ METHOD
Chemical properties	Attenuated total reflectance- Fourier transform Infrared (ATR-FTIR) spectroscopy	To uncover bands that can be attributed not only to the PA selective layer but also to the underlying substrate. Additionally, it is used to identify the presence of organic and/or inorganic nanofillers within the PA layer.
	Energy-dispersive X-ray spectroscopy (EDS)	To provide spatial distribution of elements and their respective concentrations by conducting a mapping of the PA layer.
	X-ray photoelectron spectroscopy (XPS)	To determine elemental composition of the PA layer and assess the cross-linking degree of PA layer.
	Zeta potential analysis	To characterize the surface charge characteristics (measured in mV) of PA TFC membranes by varying pH conditions.
	X-ray diffractometry (XRD)	To investigate the change in crystalline and amorphous properties of PA TFC membrane and to identify the presence of nanomaterials within the PA layer.

(Continued)

TABLE 2.1 (Continued) Analytical instruments/methods used to access the surface characteristics and performance of PA TFC membranes for water application

PROPERTIES ASSESSMENT	TYPE OF INSTRUMENT/ METHOD	FUNCTION(S) OF INSTRUMENT/ METHOD
Physical properties	Scanning electron microscopy (SEM)	To analyze the surface morphology of PA layer and cross-section structure of substrate.
	Atomic force microscopy (AFM)	To gain molecular-level insights into membrane morphology, and the interactions occurring between the membrane and fouling substances.
	Positron annihilation spectroscopy (PAS)	To identify the characteristics of free-volume holes within the TFC membrane and estimate the thickness of the PA layer.
	Contact angle analysis	To evaluate the wettability of the PA layer.
Stability test	Fouling	To study the effects of various types of foulants on the separation performance of the membrane.
	Chlorination	To evaluate the membrane resistance to chlorine exposure or attack
	Compaction	To evaluate the consistent performance of a membrane when subjected to increased operating pressure.
	Thermal	To assess the thermal stability of the membrane when subjected to elevated or harshly hot conditions.
Separation performance	Dead-end/crossflow filtration	To assess membrane performance parameters such as water flux and solute rejection under different testing conditions or environments.

focusing on their chemical and physical properties, separation performance and stability. Typically, prior to conducting membrane filtration experiments, a range of techniques can be applied to characterize PA TFC membranes. This is done to gain a comprehensive understanding of various membrane parameters that play a crucial role in achieving the desired balance between water flux and solute rejection when designing a membrane with improved features compared to the commercial membranes.

2.2 CHARACTERIZATION OF CHEMICAL PROPERTIES OF MEMBRANES

In this section, a brief overview of the techniques employed by researchers for the chemical characterization of PA TFC membranes is provided. One commonly used technique for assessing the chemical properties of these membranes is attenuated total reflectance-Fourier transform infrared (ATR-FTIR) spectroscopy. ATR-FTIR spectroscopy allows for the direct examination of dry membrane samples without the need for additional preparation. This technique is particularly valuable for analyzing the functional groups and chemical bonds present in PA TFC membranes that are made from various combinations of monomers and substrates.

Another important tool for characterization is energy-dispersive X-ray spectroscopy (EDS). EDS provides valuable insights into the distribution of elements, typically carbon (C), oxygen (O) and nitrogen (N), within the PA layer. It also provides information about the mass and atomic percentages of these elements. Researchers also frequently use EDS analysis to examine the distribution of nanofillers, whether embedded within the PA layer or coated on top of it, by detecting specific elements associated with these nanofillers.

In contrast to ATR-FTIR spectroscopy, which primarily offers qualitative data regarding the presence of functional groups within the PA layer, X-ray photoelectron spectroscopy (XPS) has several advantages. It provides precise quantitative information about the composition of the PA layer, with the ability to probe to a depth of just a few nm. Additionally, XPS offers deconvoluted spectra for C1s, N1s and O1s, which are highly beneficial for gaining insights into the degree of cross-linking within the PA structure.

The surface charge in PA TFC membranes plays a crucial role in filtration processes. Zeta potential is a commonly used parameter for characterizing the surface charge properties of these membranes under various pH conditions. This analysis is particularly significant for understanding the acid-base behavior of TFC membranes, predicting separation efficiency and assessing fouling tendencies at different water pH levels. In general, PA TFC membranes tend to exhibit a more positive charge at pH levels below 3.5 due to the protonation of amine functional groups. However, at higher pH values, they tend to display a negative charge because of the deprotonation of these amine functional groups. A negatively charged TFC membrane is highly desirable as it can effectively reject ions and reduce surface fouling. This information is valuable for optimizing membrane performance in various filtration applications.

2.3 CHARACTERIZATION OF PHYSICAL PROPERTIES OF MEMBRANES

In this section, a brief description of the techniques used for the physical characterization of PA TFC membranes is provided. Scanning electron microscopy (SEM) is a valuable tool for investigating the surface and cross-sectional morphology of PA TFC membranes at various magnifications. While SEM does not provide the high-resolution images of transmission electron microscopy (TEM), it offers magnifications of up to 10,000–15,000 times, which are adequate for examining the surface structure of the PA layer and discerning substrate morphology. Additionally, SEM analysis involves relatively simple and quick sample preparation. Before conducting SEM analysis, the membrane sample only requires a thin coating of materials like gold (Au) or platinum (Pt). This coating serves to prevent surface charging and reduce thermal damage, enhancing the quality of sample image during analysis.

In addition to SEM, atomic force microscopy (AFM) serves as a valuable complement for characterizing PA TFC membranes. AFM enables the generation of three-dimensional (3D) topographic images of the PA layer and facilitates the determination of surface roughness parameters such as average roughness (R_a), root mean square (RMS) roughness and surface area differences. This technique employs a sharp tip to scan the membrane sample's surface, allowing for the mapping of contours and capturing images at the atomic scale. Moreover, AFM is instrumental in studying material-membrane interactions, especially when employing a colloidal probe with microspheres to simulate foulants. This capability makes AFM a versatile tool for not only surface characterization but also for exploring how membranes interact with various substances, providing valuable insights for the filtration process.

For a comprehensive understanding of the structural properties of PA TFC membranes, positron annihilation spectroscopy (PAS) is utilized as an advanced analytical tool. PAS is exceptionally valuable for detecting molecular vacancies and pores within membrane materials, providing crucial information at the nanometer scale. Specifically, PAS has the capability to identify free-volume holes within the TFC membrane and accurately estimate the thickness of both the PA layer and the transition layer (located between the PA and the substrate). While SEM and AFM are limited to detecting static defects near the surface of the PA layer, PAS overcomes these limitations by offering insights into cavity properties at various depths (up to approximately 8 μm) within the membrane. Therefore, the use of PAS in membrane characterization provides a more comprehensive picture of the membrane's structural properties and enhances understanding of its composition and performance.

To assess the wettability of the PA layer, researchers commonly perform water contact angle measurements. An ideal superhydrophilic membrane surface would typically exhibit a water contact angle close to zero, but achieving this on a dense and rough surface can be challenging. Typically, the water contact angle of a PA TFC membrane falls within the range of 20° to 70°, and this range can vary based on the surface chemistry and morphology of the membrane. It is important to note that when a water droplet is placed on the membrane surface, the shape it forms is influenced not only by the chemical interactions between the surface and the liquid but also by the surface roughness. Therefore, to obtain a more accurate assessment of wettability, researchers often use roughness-corrected contact angles, which help account for the influence of surface roughness on the contact angle measurement. This correction is essential for obtaining a more precise understanding of how the membrane interacts with liquids.

2.4 SEPARATION PERFORMANCE EVALUATION OF MEMBRANES

The two primary parameters crucial for assessing the filtration efficiency of PA TFC membranes are water flux and solute rejection. These parameters can be determined through filtration experiments which are typically conducted using either a dead-end filtration setup or a cross-flow filtration system. However, for laboratory-scale experiments, the dead-end filtration mode is often preferred due to its simplicity in terms of both experimental setup and operation. It is important to emphasize that while the dead-end filtration setup is advantageous for lab-scale experiments, it is not suitable for prolonged experimental durations. This limitation arises from the significant accumulation of solutes near the membrane surface over time, which can substantially interfere with the accurate measurement of water permeability.

The literature on PA TFC membranes often employs a variety of units to express water flux, which can lead to challenges when comparing research findings. These units used include L m^{-2} h^{-1} (LMH), gal ft^{-2} day^{-1}, kg m^{-2} h^{-1}, m^3 m^{-2} h^{-1}, m s^{-1} and cm^3 cm^{-2} s^{-1}. Due to the absence of a standardized method for determining membrane water flux, researchers may encounter difficulties when evaluating membrane performance reported by different sources. Moreover, the use of different operating pressures further complicates the direct comparison of membrane performance. To mitigate this issue, many researchers opt to report water flux in the unit of L m^{-2} h^{-1} (LMH) and water permeability in L m^{-2} h^{-1} bar^{-1} (LMH/bar) This approach helps establish a more consistent and comparable framework for assessing PA TFC membrane performance across various studies.

The rejection capacity of PA TFC membranes, especially in the NF category, is influenced by two distinct mechanisms, namely, the sieving effect and the Donnan exclusion effect. To comprehensively evaluate the membrane's performance, researchers commonly employ two different types of test solutes, i.e., neutral solutes (e.g., polyethylene glycols) with molecular weights (MW) falling within the range of 100 to 1000 g mol^{-1} and charged solutes with varying charge number. This dual approach is especially relevant in applications where separation of both neutral and charged species is essential, such as in water treatment and ion-selective filtration processes.

For PA TFC membranes designed specifically for RO processes, the standard test solute used to assess membrane separation efficiency is typically sodium chloride (NaCl). Researchers conducting laboratory experiments commonly prepare NaCl solutions with concentrations in the range of 500–2000 mg L^{-1}. The rationale behind selecting this concentration range is to avoid the substantial osmotic pressure that could significantly influence water transport through the membrane. However, it is worth noting that some studies may evaluate the performance of RO membranes under conditions that align with the standard testing protocols applied by membrane manufacturers. For seawater desalination, membrane manufacturers might use a NaCl solution with a concentration of 32,000 mg L^{-1} at an operating pressure of 55 bar to simulate seawater desalination processes. This condition is representative of the high salinity of seawater. For brackish water desalination, the standard testing conditions may involve a NaCl solution with a concentration of 2,000 mg L^{-1} at an operating pressure of 15.5 bar. Brackish water has a lower salt content compared to seawater, and these conditions are more appropriate for this type of feedwater. These standard testing conditions are designed to mimic the challenges faced in practical RO applications, such as desalination of seawater and brackish water. They provide a benchmark for assessing the performance of TFC RO membranes under conditions that reflect their intended use, making it easier for researchers to compare and evaluate different membrane products for specific applications.

2.5 STABILITY TEST OF MEMBRANES

The performance stability of PA TFC membranes is heavily influenced by the characteristics of the feed solution. Filtration processes involving TFC membranes commonly encounter two significant issues, namely, surface fouling and chlorination. Surface fouling of a TFC membrane refers to the undesirable accumulation and deposition of various substances on the PA layer. This fouling can arise from different sources, including organic compounds, inorganic salts and biological growth. Chlorination, on the other hand, involves the degradation of the PA structure when the membrane comes into contact with chlorine, a commonly used disinfectant in water and wastewater treatment.

It is important to note that there is no universally standardized approach for conducting fouling and chlorination tests, and methodologies can vary significantly between studies. However, some typical practices among most studies include (1) utilizing model foulants like bovine serum albumin (BSA) and sodium alginate (NaAlg) at concentrations ranging from several hundred mg L^{-1} to one thousand mg L^{-1} as the feed for fouling tests; and (2) employing sodium hypochlorite (NaOCl) solutions containing active chlorine at predetermined concentrations, often ranging from 100 to 2500 mg L^{-1} and subjecting the membrane to this chlorinated solution for a specified duration. In both stability tests, the performance of the membrane is continuously monitored over time in terms of water flux and solute rejection. This assessment typically spans a duration of hours to days, enabling researchers to observe and analyze how fouling or chlorination influences the membrane performance over time.

Similar to the evaluation of membrane stability against fouling and chlorination, the impacts of compaction and thermal stress on membrane performance are investigated over an extended filtration period. This evaluation involves monitoring changes in water flux, solute rejection, as well as structural and surface chemistry alterations. In the compaction test for a TFC membrane used in water filtration, the goal is to assess the membrane resistance, including both the PA layer and the microporous substrate, as well as physical deformation or compaction caused by applied pressure. This test serves to determine the membrane mechanical stability and its capacity to withstand the conditions it may encounter during the filtration processes. For the thermal stability test, the TFC membrane is assessed for its resistance to high-temperature feed solutions. This is achieved by investigating both the membrane separation efficiency and its structural integrity under elevated temperature conditions.

In both cases, monitoring water flux, solute rejection, structural changes and surface chemistry alterations over an extended period allows researchers to gain a comprehensive understanding of the membrane durability in harsh operating conditions. This information is valuable for designing and selecting membranes that can reliably perform in the targeted water filtration applications.

Advanced Approaches for Polyamide Layer Characterization

3

3.1 NUCLEAR MAGNETIC RESONANCE

The utilization of nuclear magnetic resonance (NMR) spectroscopy in assessing the inherent properties of PA layer is relatively infrequent. Nevertheless, studies have indicated the potential of NMR spectroscopy in detecting alterations in the cross-linking degree of the PA layer following chlorine exposure [6], as well as in analyzing the chemical structure of PA layer comprising newly synthesized monomers or copolymers [7]. For NMR analysis, the membrane sample has to be cut into small pieces and dissolved in strong solvents such as deuterated chloroform [8] and dimethyl sulfoxide [6]. After that, the dissolved sample (in the NMR tube) is placed between two poles of a powerful magnet. The magnetic radio waves from the NMR are then transmitted to the sample, exciting the sample's atomic nuclei. Subsequently, the produced resonance frequency from the atomic nuclei is detected by a radio receiver [9].

Using NMR spectroscopy, La et al. [6] compared the ^{1}H NMR spectra of a conventional TFC membrane made of MPD and TMC monomers (designated as Ref-PA) with their newly developed chlorine resistant hexafluoroalcohol (HFA)-TMC (HFA-PA) membrane before and after exposure to 500-ppm hypochlorous acid. The researchers elucidated that the steric and electron-withdrawing characteristic of HFA could effectively minimize chlorine attack on the amide group and benzene ring of the polymer backbone, which led to enhanced membrane durability during desalination process. Analysis of

DOI: 10.1201/9781032690346-3

the NMR spectra revealed the emergence of multiple fresh peaks within the Ref-PA membrane subsequent to chlorine exposure. These peaks indicated a predominant substitution of chlorine molecules onto the benzene ring, specifically at the "e" and "f" positions of the cross-linked PA structure, attributed to Orton rearrangement resulting from N-chlorination. This finding confirmed the degradation of the Ref-PA membrane due to the impact of chlorine attack. In contrast, the HFA-PA membrane exhibited exceptional resistance to chlorine, as evidenced by minimal alterations observed in its NMR spectrum. These findings aligned with the separation efficiency evaluation, in which both membranes were soaked in 1000-ppm HOCl solution for 24 h followed by NaCl rejection test using a feed solution of pH 9.7 at 27.6 bar. The normalized salt rejection of the Ref-PA membrane decreased to below 0.30, while the HFA-PA membrane sustained a rejection close to 1.00. These results affirmed the beneficial impact of HFA in fortifying the PA layer against chlorine-induced deterioration.

Sun et al. [10] also employed the NMR technique to determine the chlorine tolerance of their lab-synthesized TFC membranes. They coated the PA surface with a thin layer made of terpolymer, i.e., 2-acrylamido-2-methyl propane sulfonic acid, acrylamide and 1-vinylimidazole (P(AMPS-co-Am-co-VI)). Prior to the coating, the authors used NMR spectroscopy to analyze the intrinsic properties of terpolymer, and the ^{1}H NMR spectrum showed a significant peak in the region of 4.5–4.0 and 2.5–1.5 ppm, which corresponded to –CH$_2$– and –CH– in the polymer backbone, respectively. The AMPS contents could be observed at 3.1 ppm (–CH$_2$–) and 1.5 ppm (–CH$_3$–), while the protons of imidazole ring (VI) were detected at 7.4 ppm and 8.6 ppm. In addition, the authors were able to confirm the monomer ratio of AM:AMPS:VI (3:3:2) in the terpolymer based on the integration of the NMR peaks. Utilizing the same technique, it was noted that the chlorine resistance of the P(AMPS-co-Am-co-VI)-coated TFC membrane superior performance when compared to the control TFC membrane. In this work, both membranes were analyzed after undergoing 40-h exposure to the 2000-ppm NaClO solution at pH 4. Notably, there was no considerable decrease observed in the hydrogen intensity of the coated membrane. This outcome was attributed to the existence of an additional layer, which functioned as a barrier and thereby mitigated direct chlorine attack on the PA matrix. Furthermore, the coated membrane also demonstrated excellent long-term performance stability with respect to flux and salt rejection after chlorine exposure to 80,000 ppm.h for up to 15 days. The durability of the coating layer is likely due to the increased π-π stacking and hydrogen bonds caused by VI and AM in the terpolymer.

Rather than employing ^{1}H NMR, Almijbilee et al. [11] opted for ^{13}C NMR to analyze the organic structure of the cross-linked PA layer, composed of 1% 4,4′-oxydianiline (ODA) and 0.2% TMC monomers. The primary distinction between ^{13}C NMR and ^{1}H NMR lies in their respective functions, specifically in that ^{13}C NMR is utilized to identify the quantity and nature of carbon atoms within a molecule, whereas ^{1}H NMR is employed to ascertain the quantity and nature of hydrogen atoms present in a molecule [12]. Since the aromatic rings of ODA can form π-π stacking interaction with the aromatic section in the backbone of the C-PEI membrane substrate, the existence of the aromatic ring (attached to amine) could be detected by NMR at 146.5 ppm. Additionally, the peaks belonging to phenyl rings (attached with oxygen) and carboxylic acid could be detected at 155 ppm and 172 ppm, respectively.

Despite the capability of amine monomers like PIP and MPD to develop high-rejection TFC membranes through cross-linking with TMC, their performance is always constrained by the trade-off between water permeability and selectivity [13]. In order to alleviate this issue, Kazemi et al. [14] utilized monomers with three amino groups, namely, 1,2,4-triaminobenzene (TAB), which was synthesized from 2-nitro-1,4-phenylenediamine precursor in the presence of palladium on carbon (Pd/C) catalyst (Figure 3.1(a)). The authors varied the ratio between TAB and MPD during the IP process to develop a new type of PA structure, as illustrated in Figure 3.1(b). The molecular structure of TAB was subsequently confirmed through ^{1}H NMR spectroscopy, as depicted in Figure 3.1(c). The peaks associated with TAB (labelled as A, B and C) were observed at higher chemical shift values (7.367, 6.813 and 6.979 ppm, respectively) in comparison to the peaks of 2-nitro-1,4-m-phenylenediamine (which appeared at 6.579, 6.161 and 6.247 ppm). The chemical shift in the TAB can be attributed to the reduction of the nitro group to an amine group. Despite showing enhanced water flux, the salt rejection of the modified TFC membrane plummeted from 97.7% to 77.5% as the TAB concentration increased from 0.3% to 2.0 w/v%. The authors attributed the reduced salt rejection to the thinner PA film formed that resulted from the slow diffusion rate of TAB toward the TMC organic phase.

Using 4-(piperazin-1-yl)benzene-1,3-diamine (PMPD) to react with TMC, Li et al. [15] successfully developed a new type of TFC membrane for NF. The authors synthesized PMPD by combining 4-fluoronitrobenzene and Boc-protected PIP. The Boc-group was then de-protected under acid condition after the aromatic ring nucleophilic substitution reaction. Following this, Pd-catalyzed reduction reaction was applied to convert the monomer into PMPD. The purity of the self-synthesized PMPD was confirmed through NMR analysis, as depicted in Figure 3.2. According to the ^{1}H NMR analysis, the protons originating from the methylene groups of the PIP unit were identified at 2.95 and 2.76 ppm. Additionally, peaks corresponding to protons e, d and c of the benzene ring were observed at 6.23, 6.15 and 6.12 ppm, respectively. Meanwhile, the two aliphatic carbon and six phenyl carbon atoms of the monomer could also be identified in the ^{13}C NMR spectra. Regarding

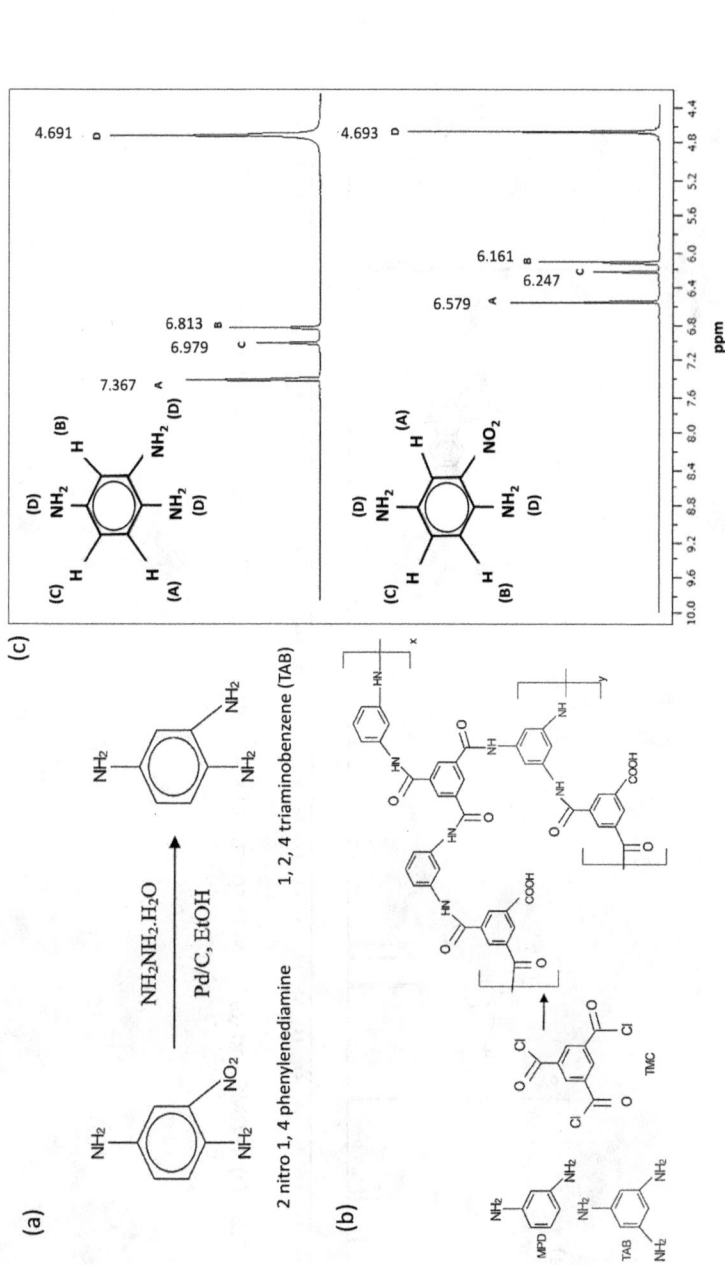

FIGURE 3.1 (a) Synthesis of 1,2,4-triaminobenzene (TAB) from 2-nitro-1,4-phenylenediamine, (b) Synthesis of PA network using MPD, TAB and TMC via IP process and (c) ¹H NMR analysis for TAB and 2-nitro-m-phenylenediamine [14].

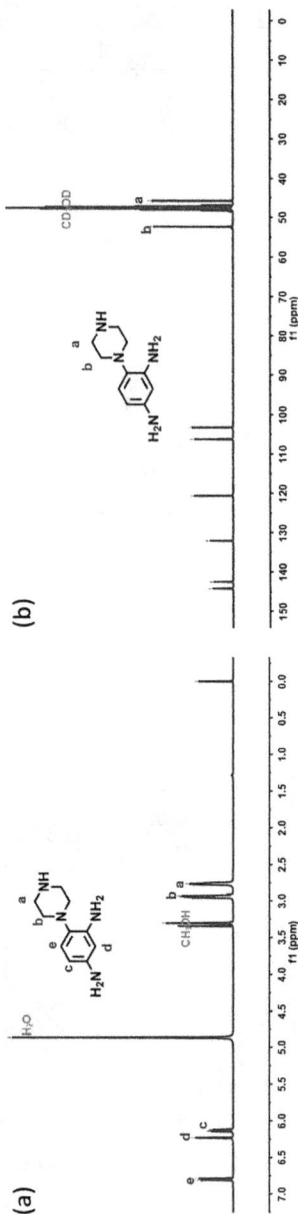

FIGURE 3.2   (a) ¹H NMR and (b) ¹³C NMR spectra of self-synthesized PMPD (inset: Organic structure of PMPD) [15].

membrane performance, the presence of 0.80 wt% PMPD in the membrane led to improved performance in both pure water permeability and Na_2SO_4 rejection. There was an increase of approximately 66.7% in pure water permeability and around 8% enhancement in Na_2SO_4 rejection compared to the membrane modified with a lower concentration of PMPD (0.10 wt%). This enhancement in performance could be attributed to a higher degree of cross-linking within the PA layer combined with a larger MWCO.

Compared to Raman spectroscopy, NMR spectroscopy can detect polar compounds and its spectrum has less background interference. Moreover, it is an easily quantifiable technique and permits the routine identification of novel compounds [16]. Nevertheless, NMR analysis comes with several limitations, among which the most prominent challenge is its limited sensitivity. The inherently low signal intensity is primarily attributed to the low magnetic moment of atomic nuclei [16, 17]. This low signal intensity poses a significant limitation. For instance, elements with paramagnetic properties, like oxygen (O), exhibit fewer NMR signals compared to hydrogen (H) and carbon (C) atoms due to strong hyperfine interactions, consequently reducing the precision of atomic measurements [18].

3.2 RUTHERFORD BACKSCATTERING SPECTROMETRY

Rutherford backscattering spectrometry (RBS) is a powerful quantitative characterization technique used for analyzing membrane surfaces. Its strength lies in its ability to assess elemental depth heterogeneity, allowing for the determination of layer thicknesses and elemental compositions within the membrane [19]. Additionally, RBS has the capability to evaluate the porosity of the substrate [20], as well as assess parameters such as charge density and cross-linking degree within the PA layer [21]. Generally, the depth precision of RBS ranges from a few µm to around 20–30 nm. As a result, RBS is most effective in discerning the inherent characteristics of the extremely thin PA layer in TFC membranes, especially when compared to their significantly thicker support layers.

RBS is capable of quantifying both the quantity and energy levels of ions that are deflected backward from a solid specimen. This is achieved by employing a mono-energetic beam of light ions, commonly MeV helium (He) ions, as the incoming probing source. The RBS technique utilizes a silicon (Si) surface barrier detector placed at a specific backscattered angle in relation to the ion beam. This detector is employed to analyze the scattering of He particles emitted from the target sample, commonly falling within the range of 100° to 170°. The energy of the backscattered He ions are then inferred by the conservation of momentum and energy between the scattering atom and incident ion, as well as the mass and depth of the target sample. Ultimately, the elemental concentration can be deduced since it correlates directly with the quantity of ions that undergo backscattering [22].

To assess the precision of established measurement methods, Lin et al. [23] utilized a variety of approaches, such as SEM, TEM, AFM, profilometer, ellipsometer, quartz crystal microbalance (QCM) and RBS, to gauge the thickness of the PA layer in six different commercial TFC NF/RO membranes. The findings of their study indicated that among the diverse techniques, RBS exhibited the smallest average standard deviation in measuring the PA layer thickness (4%), outperforming the other methods that yielded deviations ranging from 10% to 18%. This indicated the PA layer thickness measured via RBS is more likely to be representative of the actual PA layer of TFC membrane. The analysis also revealed distinct differences in the RBS spectra between the polysulfone (PSf) substrate and the PA layer of the membrane. This differentiation was based on the presence of peaks corresponding to sulfur (S) and nitrogen (N) elements, whereby S was detected solely in the substrate, while N was exclusively identified in the PA layer.

Gorzalski et al. [19] performed depth-homogeneous and depth-heterogeneous analysis on the PSf substrate and TFC membranes using three different analytical techniques, namely RBS, energy-dispersive X-ray spectroscopy (EDS) and X-ray photoelectron spectroscopy (XPS). The depth-homogeneous results indicated that the elemental composition of PSf substrate analyzed by EDS yielded lower C but higher S and O contents ($C_{0.768}O_{0.179}S_{0.053}$) than XPS ($C_{0.861}O_{0.112}S_{0.027}$) and RBS ($C_{0.845}O_{0.124}S_{0.031}$). In contrast with the theoretical PSf composition ($C_{0.844}O_{0.125}S_{0.031}$), a comparison revealed that the elemental composition derived through RBS analysis exhibited an exceedingly minimal variance of less than 1%. This clearly demonstrated the high accuracy of RBS in ascertaining surface chemistry. For TFC membrane characterization, the authors performed a depth-heterogeneous analysis to focus only on the PA layer and the results are shown in Figure 3.3. Utilizing EDS with depth penetration, the elemental composition obtained closely resembled that of the PSf layer. Contrary to the EDS, the RBS with nm-scale depth resolution was able to resolve the PA layer (with main element of C, N and O) from the PSf substrate, offering results similar to that of XPS analysis.

Considering the varied capabilities and limitations such as differing analysis depths associated with XPS, EDS and RBS techniques, Gorzalski et al. [19] conducted a study to discern the strengths and weaknesses of employing these characterization methods. Their study focused on analyzing the fouling

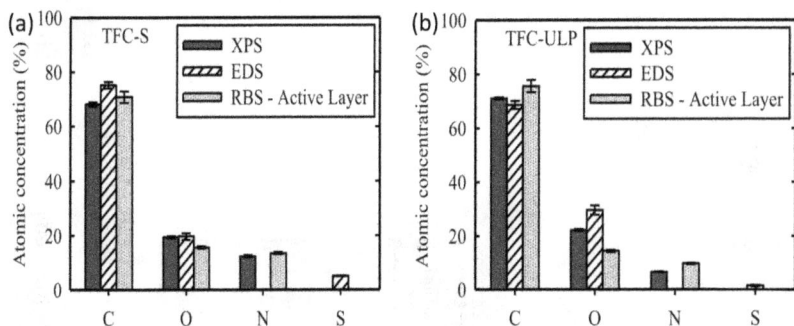

FIGURE 3.3 Element compositions of (a) TFC-S and (b) TFC-UKP membranes measured based on XPS, EDS and RBS [19].

Note: TFC-S and TFC-ULP membranes are the membranes manufactured by Koch Membrane Systems.

properties of TFC membranes both before and after undergoing chemical cleaning. Their findings revealed the inadequacies of relying solely on a single technique for comprehending the composition and structure of foulant layers. The limitations inherent in each method could lead to incomplete or inaccurate conclusions. For instance, EDS may fall short in detecting minute traces of foulant, while XPS may be unable to identify major elements in samples with depth heterogeneity. Although RBS stands out as a superior approach for elemental analysis, it lacks the heightened sensitivity exhibited by XPS when it comes to detecting trace amounts of foulant.

Utilizing the RBS technique, Valentino et al. [24] successfully identified and delineated the impact of monochloramine, as well as bromine and iodine-containing secondary oxidizing agents, on the physicochemical properties of the PA layer. Monochloramine is a disinfectant and could react with bromide and iodide, which are typically present in seawater at concentrations of 65 mg/L and 60 μg/L, respectively. Figure 3.4(a) illustrates the interaction of PA layer with the oxidized bromine species (HOBr or Br_2) and the change in organic structure that could deteriorate the membrane separation performance. Furthermore, Figure 3.4(b) presents the level of bromination of the PA layer based on RBS analysis. The investigation revealed a noticeable rise in the bromination of the PA layer from 0.26% to 1.10%. This increase correlated with the monochloramine exposure duration, which progressed from 0.24 h to 1.92 h. This phenomenon could potentially be attributed to the assimilation of halogens caused by the generation of secondary oxidants resulting from the interaction between bromide and monochloramine.

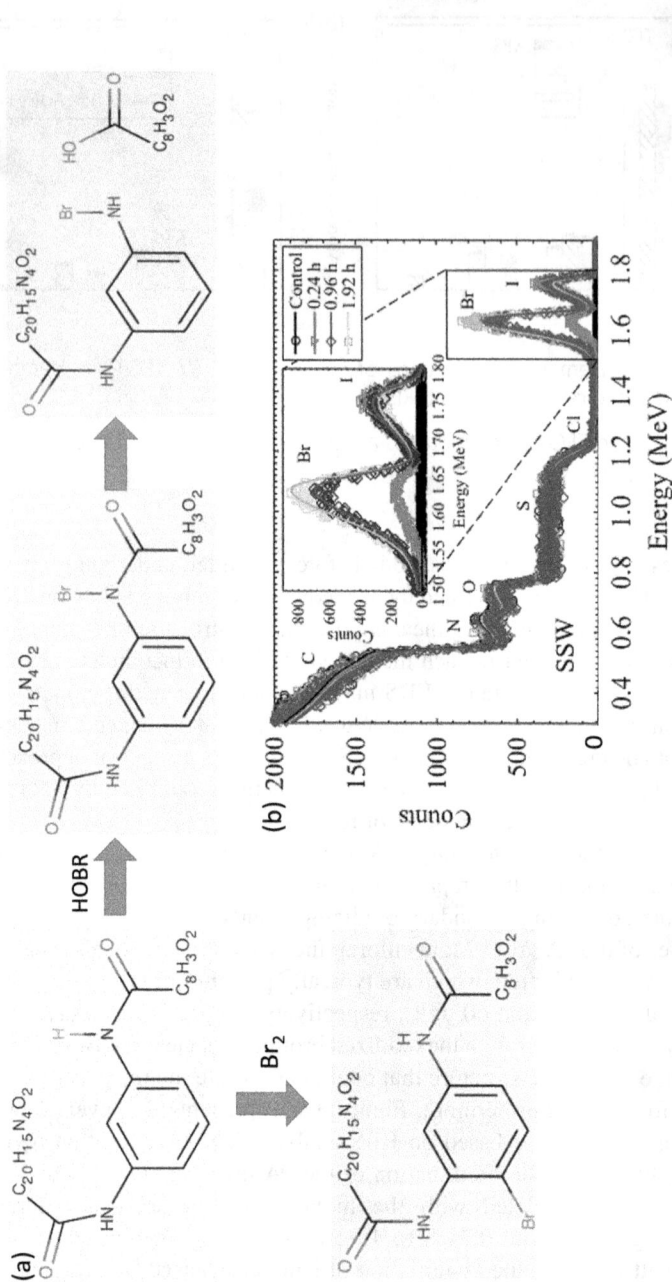

FIGURE 3.4   (a) Chemical structure of PA layer before and after bromination and (b) RBS spectra of RO membrane samples after being exposed to chloraminated synthetic seawater [24].

While RBS presents advantages such as quantitative and non-destructive surface analysis, along with the absence of a need for reference standards, it does exhibit certain limitations. Notably, its sensitivity toward lighter elements is inferior to that for heavier elements. Furthermore, RBS is inadequate in distinguishing between two or more elements possessing similar atomic numbers, largely due to constraints in mass resolution and/or potential spectral overlap with depth. Additionally, RBS is a rather complex technique that requires specialized equipment that is not commonly available.

3.3 WIDE-ANGLE AND SMALL-ANGLE X-RAY SCATTERING

In parallel to the efforts of direct imaging techniques, progress has also been made for indirect non-imaging techniques. The indirect methods such as small-angle X-ray scattering (SAXS) and wide-angle X-ray scattering (WAXS) are advanced analytical tools that measure the intensities of X-rays scattered in a sample as a function of the scattering angle. The monochromatic X-rays are transmitted to the sample, and the scattered X-rays are detected and analyzed to study the characteristics of the particles in the system, as illustrated in Figure 3.5.

Figure 3.6 shows an 1D scattering intensity profile of PA by integrating a 2D SAXS pattern [26]. The information obtainable include (i) particle size, (ii) pore size distribution, (iii) chemical composition and (iv) polymeric chain structure formation in the PA chain. In the Guinier region, scattering intensity, I, is proportional to the concentration of the system and thus independent of q.

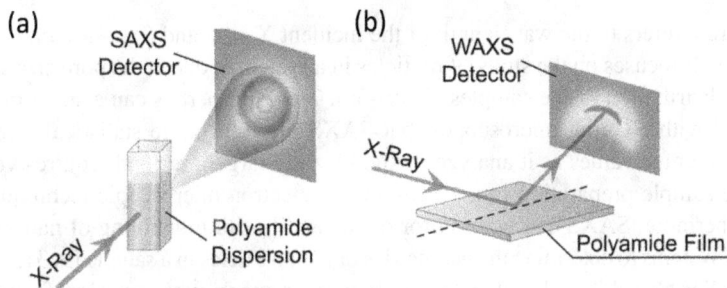

FIGURE 3.5 Schematic illustration of (a) SAXS PA bulk dispersion where solid PA film is ground into small flakes and dispersed in water as colloidal-like suspension, resulting in isotropic ring patterns on a 2D SAXS detector and (b) WAXS analysis on PA film [25].

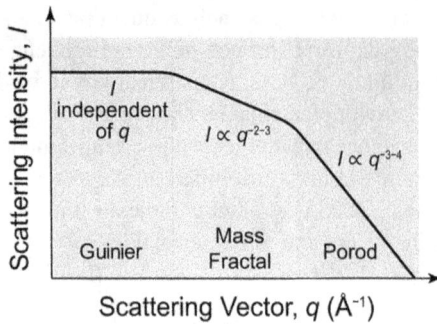

FIGURE 3.6 Schematic diagram of 1D scattering intensity profile of PA by integrating a 2D SAXS pattern [26].

In the mass fractal ($I \propto q^{-2-3}$) region, the correlation of I with q provides structural information of the large clusters of the PA chain. In the Porod ($I \propto q^{-3-4}$) region, the correlation of I with q provides the structural information of the small primary units of the PA chain [25]. These information are valuable to thoroughly understand the internal structures of TFC membranes. Additionally, this non-destructive analysis is beneficial for revealing the preferential orientation (i.e., in-plane or out-of-plane configurations) and molecular structure of the PA layer [25, 27].

In principle, SAXS measures the intensities of the X-rays scattered by a sample as a function of scattering angles less than 5°, with a wide scattering wave vector q-range of 10^{-3} nm^{-1} < q < 8 nm^{-1}:

$$q = \left(\frac{4\pi}{\lambda}\right)\sin\left(\frac{\theta}{2}\right)$$

(3.1)

where λ refers to the wavelength of the incident X-rays and θ *is the scattering angle*. It focuses on the study of particles in a system to detect the pore size and size distribution of the samples [9, 28–30]. These properties can also be measured with electron microscopes, but SAXS provides more statistically reliable average values as it analyzes a significantly larger area and requires very little sample preparation time in contrast to electron microscopic techniques. Furthermore, SAXS can be used for online and in-situ monitoring of nanoparticle systems to determine the geometries of nanoparticles in a sample [9, 31, 32]. SAXS is also able to detect and quantify measurements with structural features in the nm length scale, typically between 1 and 100 nm.

WAXS functions similar to SAXS, with a key difference being the scattering angle range. The increased angle (>5°) between the detector and the sample for WAXS allows the probing of material structures in the interatomic range and the scattering intensity is plotted as a function of the 2θ angle. A typical WAXS detector covers a much wider q-range of 5 nm^{-1} < q < 50 nm^{-1}. WAXS is often used to determine the degree of crystallinity of polymer texture and preferred alignment of crystallites of a film, chemical or phase composition, presence of film stress and crystallite size to obtain information on the structural effects on performance. In brief, WAXS at large scattering angles probes atomic/molecular length scales, while SAXS at intermediate and small scattering angles probes nm to millimeter (mm) length scales.

In 2020, a study was carried out to compare different techniques namely, SAXS, WAXS and X-ray diffraction (XRD) in analyzing the structure, shape and pore size of PA TFC membranes [9]. The authors deemed XRD a less versatile tool compared to SAXS and WAXS. The latter tools are preferred over the former as X-ray scattering can be performed on all types of membranes with various structures, such as non-crystalline, semi-crystalline and crystalline structures, while XRD application is limited to large crystalline structures. Furthermore, it was reported that the overlap of XRD peaks tends to occur when multiple crystalline compounds or phases are present within a sample. This overlap can complicate the analysis and interpretation of the diffraction pattern obtained from the sample. The findings revealed SAXS as a better tool to determine the nm-scale pore size and distribution for larger areas, while WAXS scans to study the chemical composition and degree of crystallinity. As technologies advance, the duration of analysis is a considerable factor to many. For this, SAXS is at an advantage as it can statistically collect data of polymers with different cross-linked structures within seconds.

Liu et al. [33] employed SAXS as a tool to analyze the physicochemical properties of layer-by-layer (LBL) PA films. These films were created using varying ratios of MPD and TMC to explore characteristics such as surface roughness and degree of crosslinking. SAXS quantified the average molecular spacing of the films through the scattered intensity, $I(Q)$, and scattered vector, Q, obtained. With that, a correlation between concentration ratio of monomers and aperture size of PA film can be drawn. Two mLBL membranes with different MPD/TMC ratios that underwent 15 deposition cycles were fabricated prior to SAXS analysis. As presented in Figure 3.7(a), the scatter peak of the PA film with a 2.5 MPD/TMC ratio had a higher intensity and d-spacing than the PA film with a 1.25 MPD/TMC ratio. This indicated that the former PA film had a lower average molecular spacing (3.7 Å) than the latter lower-ratio PA film (3.9 Å) owing to a higher crosslinking degree and surface roughness. Moreover, the mLBL film possessing a 2.5 ratio of MPD/TMC displayed a thickness twice that of the 1.25 MPD/TMC ratio PA film, as shown in Figure 3.7(b). This indicated that a higher MPD/TMC ratio resulted in the creation of a more cohesive, denser and thicker PA active layer, achieving a faster formation rate.

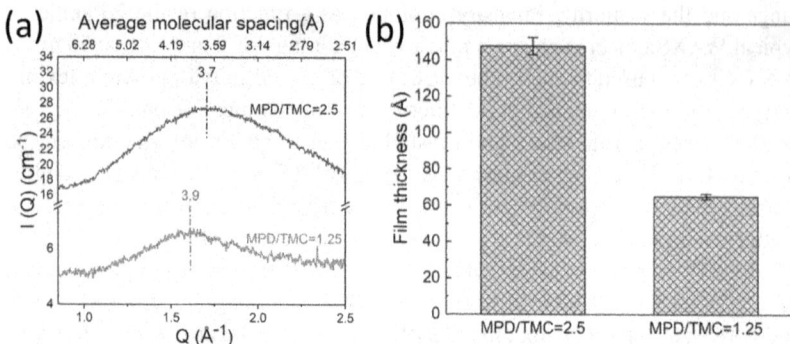

FIGURE 3.7 (a) X-ray scattering spectrum of 2.5 MPD/TMC ratio PA film and 1.25 MPD/TMC ratio PA film and (b) Film thickness of mLBL PA film possessing a different MPD/TMC [33].

In a study to evaluate polymeric molecular order [30], the network structure of a PA layer was characterized with SAXS, which was able to measure the d-spacing of the PA layer in multiple RO membranes that were prepared at different TMC concentrations. A clear shift in Q correlation peak position from 1.22 Å^{-1} to 1.16 Å^{-1} was observed. Similar characterization was also carried out on the NF membranes for comparison. Based on the results obtained from both types of membranes, SAXS was able to distinguish the minute differences between the NF and RO membranes, demonstrating that the concentration of diamine monomer reactant would result in a lower degree of polymer chain compaction and clustering. The usage of this tool in PA layer characterization provides great insights to thoroughly understand the cross-linking structure in PA layers of NF and RO membranes.

In another study [34], SAXS was used to characterize the disordered deposition of colloidal inorganic materials, namely, NaCl and MgCl$_2$, on the TFC membrane. The PA active layer was isolated and examined at minute scattering angles of less than 4° with a wavelength of 0.1542 nm using a Bonse-Hart diffractometer with slit collimation. The $q^{-2.5}$ value, which corresponds to the inverse of the atomic spacing, was below 10^{-2} Å^{-1}, upholding consistency with the mass fractal behavior.

Figure 3.8 compares the WAXS scattering pattern of a lab-synthesized 18-nm thick free-standing PA film with a PA layer obtained from a commercial membrane [26]. The horizontal (q_r) axis in the figure corresponds to the surface-parallel scattering wave vectors and the vertical (q_z) axis indicates the normal scattering wave vectors. The patterns provide details on the preferential orientation and molecular structure. It was observed that there was arc-like scattering features for the 18-nm thick free-standing PA film where intense

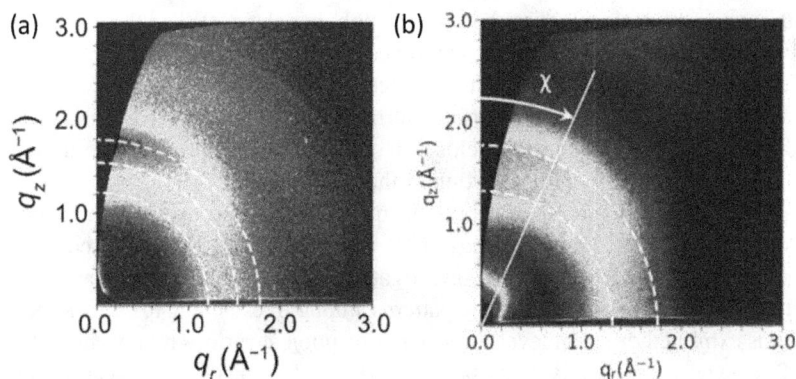

FIGURE 3.8 (a) WAXS scattering pattern of a prepared 18-nm thick free-standing PA film after citric acid post-treatment and (b) WAXS scattering pattern for a PA barrier layer obtained from the commercial membrane on a silicon substrate [26].

scattering was preferentially aligned along the q_z surface-normal direction and a weaker scattering along the q_r surface-parallel direction. The dashed arcs outlines were scattering vectors of $1.22\,\text{Å}^{-1}$, $1.54\,\text{Å}^{-1}$ and $1.79\,\text{Å}^{-1}$, corresponding to the spacings of 5.2 Å, 4.1 Å and 3.5 Å, respectively. In comparison, the commercial membrane exhibited similar arc-like scattering features as the prepared free-standing membrane, but with lower color intensity. The use of the WAXS tool for PA characterization to identify similar structural motifs of both PA films is able to efficiently aid in the validation of PA layer formation, which is useful for the advancement of membrane development studies [26].

On the other hand, Fu et al. [35] employed the WAXS technique to characterize four sets of PA active films, namely, (a) bulk powdered sample, (b) 20-nm free-standing PA film at oil/water interface, (c) 20-nm molecular LBL film and (d) 300-nm PA active layer separated from a commercial membrane. Results revealed that all PA active films exhibited broad scattering rings centered at approximately $1.67\,\text{Å}^{-1}$. The observed broad ring widths suggested that the PA layer was an amorphous phase with short-range order comparable to interatomic distances. For the bulk sample characterization, it was observed that the scattering was uniform and have an independent relationship with the angle of the surface-normal. This observation is expected as bulk powdered sample does not have preferred orientations. For the other three PA samples, the rings of scattering were more intense along q_z (surface-normal direction) while weaker scattering was observed along q_r (surface-parallel direction). Closer examination of the results revealed that the arc-scattering features did not form a uniform circle, indicating that the role of surfaces only partially aligned to the molecular backbone of the polymer chains. When the freestanding PA film was further evaluated at three tested angles (15°, 50° and 80°) with respect to surface-normal, it was reported that the atomic distance decreased as the angle with respect to surface-normal increased. This clearly indicated that the molecular spacing was closer in orientations along the surface-normal direction than the in-plane direction of the surface.

Aside from obtaining in-depth information on the core spacing, the effects of humidity on the characteristics of PA films could also be conducted using X-ray scattering. Fu et al. [35] investigated the molecular structure and conducted a comparison of scattering patterns among freestanding PA films. These films exhibited different levels of water adsorption at various relative humidity (RH) levels, specifically 46%, 60% and 100%. This analysis was carried out utilizing a humidity WAXS sample. The WAXS findings revealed that there was no significant collapse observed in the overall membrane structure when subjected to cycles of high to low humidity. Furthermore, the scattering peak closely resembled the Lorentzian profiles.

Another area that X-rays scattering tools can be extensively utilized is the application to water desalination technologies. Fouling poses a significant challenge in pressure-driven RO and NF membrane processes, leading to reduced desalination efficiency and increased energy input requirements [36]. Thus, to mitigate fouling tendencies and enhance desalination performance, it is vital to understand the growth of PA film during IP [33]. By applying X-ray scattering approaches to obtain direct information of membrane fouling, the membrane structure and chemistry can be analyzed in almost any environment. Being able to probe down to a few Ångstroms (Å) radii and operate in real water source, observations of surface-specific modality for molecular-scale characterization and synergistic reactions among foulants can be easily drawn to provide spatially resolved chemical information on the fouling area.

As SAXS and WAXS could be utilized in-situ and operando, studying the behavior of multiple foulants in the complex nature of water composition in combination with various flow profiles make it easier to predict the fouling behavior and thus the separation performance. Constituents such as organic matters and inorganic deposits are often present in a variety of source waters, especially in wastewater [36]. Due to the extensive interactions among various organic matters, it is rather challenging to understand and predict the fouling behavior in real systems without probing the chemical speciation directly [37]. The usage of X-ray techniques therefore can elucidate the chemical composition and structure of the fouling area that has such constituents present and provide detailed analysis of the molecular-scale interaction with the membrane active layer.

Moreover, SAXS offers insights into the alterations within a pore as it fills with water. This facilitates an accurate characterization of the porosity and free volume in dense membranes, leading to a more comprehensive understanding of structure-transport relationships. On the other hand, WAXS has the potential to identify the phases that exist within inorganic deposits formed on the membrane during operation [36]. This information is critical for the strategic functionalization of membranes to deter multiple foulants simultaneously to improve anti-fouling as well as implement mitigation steps in real systems.

Research conducted on X-ray scattering highlights the ability to acquire valuable data concerning the active layers, offering numerous significant advantages including: (i) non-destructive preparation on materials with various structures, (ii) analysis of the internal structures of the PA active layer, (iii) improved analysis quality on the effect of possible collapsing of active layer due to a change of environmental settings and (iv) rapid data collection. Similar to other equipment, both SAXS and WAXS have limitations. A notable constraint is the implicit requirement for sample homogeneity, which holds particular significance in SAXS analysis. Any degradation or aggregation within the prepared sample can result in misinterpretations as radiation could cause sample damage and distort the data, which limits the effectiveness of this method [9, 38, 39]. Such effect is significant when nanomaterials are not evenly distributed throughout the PA layer of TFN membrane. Hence, it is advisable to employ a variety of characterization tools to ensure the reliability of the data.

3.4 IN-SITU FOURIER TRANSFORM INFRARED SPECTROSCOPY

Fourier Transform Infrared (FTIR) spectroscopy is commonly employed for examining the PA layer of TFC membranes. However, conducting in-situ FTIR analysis to monitor the growth rate of the PA film is not a widespread practice. Gaining real-time insights into the mechanism and kinetic reaction underlying the formation of the PA layer is of utmost significance. This understanding is pivotal for the advancement of a cutting-edge selective layer (made of advanced active monomers) tailored for high-performance filtration applications. At present, only a limited number of studies have recorded findings from in-situ FTIR analysis of the development of the PA layer.

Yang et al. [40] performed in-situ FTIR analysis to monitor the growth rate of the PA layer film during the IP reaction in real time. Firstly, the substrate was fixed in a custom-made reactor to ensure that only the upper side of substrate was contacted with the PIP and TMC monomer solutions. Prior to IP, the FTIR baseline was set up in an air environment. An optic probe was then brought in contact with the substrate surface after immersing in PIP solution for 10 min. After that, the TMC solution was poured onto the PIP-saturated substrate surface and the FTIR intensity was acquired and calculated from the PA film formation process as a function of reaction time. In order to alter the IP reaction rate, the authors adopted two different strategies, i.e., incorporation of macromolecular additives (PVA) into PIP aqueous solution and establishment of ZIF-8/PEI interlayer between the PA layer and PSf substrate. Figure 3.9(a) compares the in-situ FTIR spectra of PA layer under different conditions for a reaction time ranging from 0 to 300 s. Focus was placed at the wavenumber of 1640 cm-1 as it is corresponded to carbonyl vibration of the amide (C=O). As can be clearly seen, the presence of hydrophilic ZIF-8/PEI interlayer or PVA during IP tended to slow down the IP reaction rate and reduce the PIP diffusion greatly with minimal change in the IR absorbance compared to the control sample. The authors also found that, the lower the IR absorbance was, the thinner was the PA layer, and vice versa. Specifically, the thickness of the PA layer decreased in the order of pristine PA (138 ± 24 nm) > PVA-modified PA (76 ± 19 nm) > ZIF-8/PEI-modified PA (60 ± 6 nm).

In a separate work, Yang et al. [41] acquired in-situ FTIR spectra to investigate the formation of a thin PA layer to better understand the diffusion reaction rate between hexane and the aqueous phase during IP. The researchers developed an interlayer made of tannic acid (TA) and PEI between the PA layer and substrate in order to reduce the IP reaction rate and thickness of the PA layer. As presented in Figure 3.9(b), the PA layer (corresponding to the absorbance peak at 1640 cm^{-1}) grew faster on the control substrate from 0 to 60 s compared to the modified substrate. Upon reaching the peak reaction time of 250 s, notable observations could be made. Specifically, the IR absorbance and peak area of the modified substrate exhibited a decrease, indicating a suppression in the growth rate. With regard to thickness, the PA layer that formed on both the PSf and modified PSf substrates were measured at approximately 113 nm and 75 nm, respectively.

While in-situ FTIR spectroscopy perhaps is the only characterization tool that can analyze the dynamic process of PA formation in real time and offers the capability to explore reactions across diverse solvents, pH levels and temperatures, its efficacy in pinpointing trace quantities of additives — be they organic or inorganic substances — employed to enhance PA layer attributes is limited. The presence of broad IR absorption bands can result in undesirable spectral overlap, posing challenges to quantitative analyses of complex, multi-component materials. Furthermore, it is plausible that the absorption bands observed may not solely arise from a single species, complicating the assignment of these bands to specific constituents.

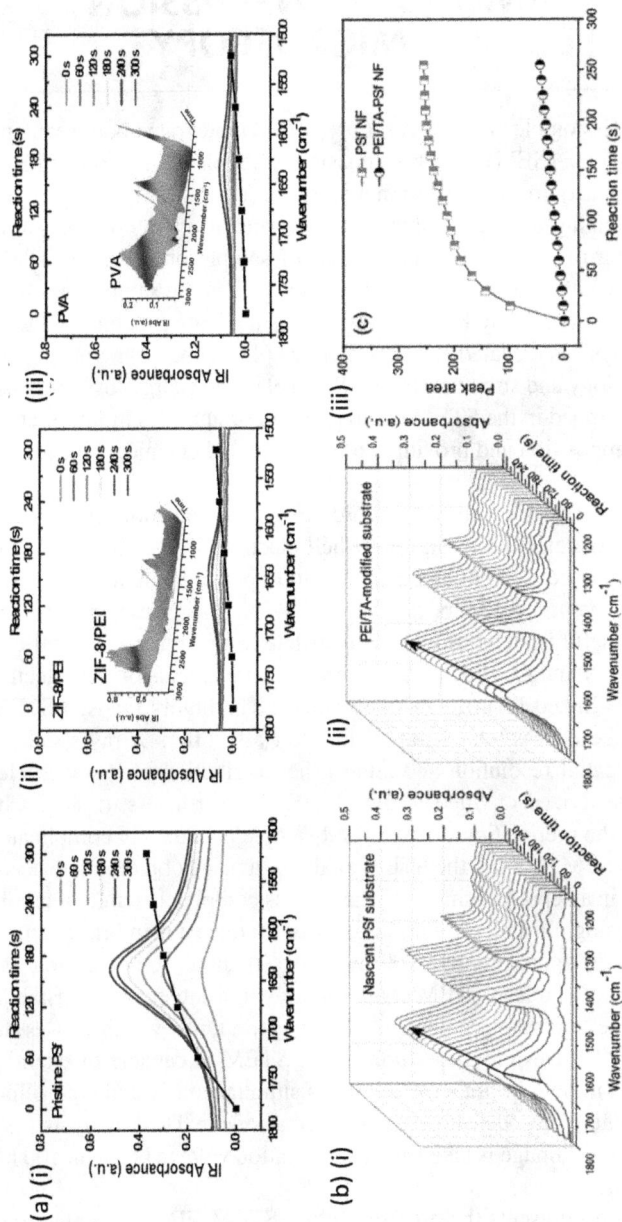

FIGURE 3.9 (a) In-situ FITR spectra of PA layer synthesized under different conditions, (i) control (PIP-TMC), (ii) PIP-TMC in the presence of ZIF-8/PEI as interlayer and (iii) PIP-TMC in the presence of PVA [40]. (b) 3D waterfall spectra of in-situ FTIR absorbance of PA formation taking place at different reaction on (i) nascent PSf substrate and (ii) PEI/TA-modified substrate and (iii) peak area at absorbance band of 1640 cm^{-1} which represents the O=C-N stretching bond of PA layer [41].

3.5 SCANNING TRANSMISSION ELECTRON MICROSCOPY

Scanning transmission electron microscopy (STEM) integrates both scanning electron microscopy (SEM) and transmission electron microscopy (TEM) methodologies. It acquires transmission images (based on TEM principles) by employing a scanning approach (akin to SEM principles). In comparison to TEM's utilization of a broad electron beam for sample observation, STEM employs a focused electron beam to systematically scan the sample in a line-by-line manner, as depicted in Figure 3.10(a). This innovative technique facilitates STEM analysis at elevated electron energies [42]. Other than viewing the material morphology and structure, the EDS and electron energy loss spectroscopy (EELS) equipped in the STEM could play important roles in interpreting the chemical composition and providing a more detailed chemical analysis on the specimen [43].

In principle, EDS is conducted according to the ionization of atoms in the specimen by the ejection of an inner shell electron. The emitted X-ray is unique to each specific element and can be used to perform chemical analysis with the help of an energy dispersive detector [44, 46]. With respect to EELS, it is extremely useful in determining the distribution of lighter chemical elements such as C, N and O, which are the three main elements of the selective layers (i.e., cross-linked PA) of TFC membranes. The highly focused STEM fitted with a probe of less than 1 nm in diameter also offers X-ray spectrum with excellent spatial resolution and allows the distribution of nanoparticles to be fully characterized or tilted in 3D [43, 47]. Nevertheless, it should be pointed out that the preparation of the STEM sample is relatively complicated compared to the SEM. Prior to the STEM analysis, a piece of membrane needs to be embedded in an epoxy resin. Once the resin is completely cured, it is subject to slicing (using ultramicrotome) in order to produce a membrane sample with a thickness of approximately 100 nm. Also, it should be noted that the serial acquisition of pixels in STEM could possibly translate position instabilities of the sample into displacements of the imaged atoms, which causes the images to blur [48]. Despite these limitations, STEM is capable of examining the specimen in both bright and dark fields simultaneously and controlling the dynamic range of the detectors to mimic that of the TEM analysis [49]. Generally, the STEM image is taken at an acceleration voltage between 100 kV and 300 kV [50].

Figure 3.9(b-c) presents the high-resolution STEM-EDS elemental mapping of the cross-section of copper (Cu)-incorporated TFC membrane before (labelled as TFC-Cu1D-3) and after (TFC-1D-3) etching by 0.2 wt% sulfuric

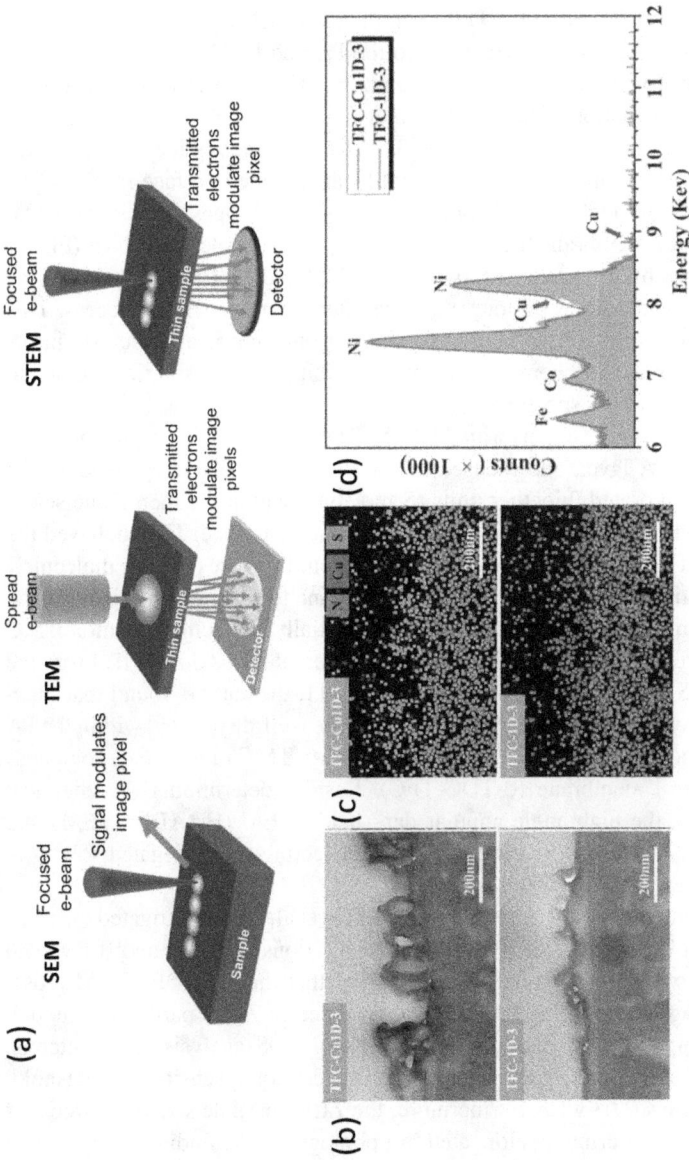

FIGURE 3.10 (a) Principle of SEM (data acquired serially), TEM (parallel image acquisition) and STEM (data acquired serially via transmitted electron) [44]. Properties of TFC membrane before (TFC-Cu1D-3) and after (TFC-1D-3) etching by 0.2 wt% H_2SO_4 solution, (b,c) STEM-EDS cross section elemental mapping and (d) EDS spectra with respect to Fe, Co, Ni and Cu [45].

acid (H_2SO_4) solution together with EDX spectra. H_2SO_4 solution was used to etch the preloaded copper hydroxide nanorods (CuNRs) into the 1D nanochannels structure. Clearly, the generation of the nanochannels inside the PA layer can be observed for the TFC-1D-3 membrane via STEM. Also, N and S components that respectively correspond to the PA and PSf layer are detected in the cross-sectional view of the TFC membrane. From the elemental mapping, significant distribution of Cu (brown color) element is delineated between the interface of PA and PSf layer of the TFC-Cu1D-3 membrane. On the other hand, there is less amount of Cu element located at the interface for the TFC-1D-3 membrane etched by H_2SO_4 solution. The changes of the elemental distribution of membrane are consistent with the EDS spectra shown (Figure 3.9(d)), whereby the peaks at 8.04 keV (CuKα) and 8.94 keV (CuKβ) of the TFC-1D-3 membrane is no longer visible after the acid-etching process. This further proved that STEM could provide not only physical structural information, but also offer chemical information such as high-resolution elemental distribution and EDS spectra.

Cao et al. [51] also performed STEM-EDS analysis to examine the properties of the PA layer, and focused on the distribution of the S element. The authors incorporated thioether units (S moieties) within the membrane selective layer with the aim of improving its chlorine resistance. They believed the oxidation of thioether units to sulfone groups could capture chlorine molecules, thus protecting the amide bonds of the membrane from being chlorinated. The authors introduced the thioether units by partially replacing common TMC with 4,4'-thiodibenzoyl chloride (TDC) in different ratios during IP. From the STEM-EDS analysis, as depicted in Figure 3.11, the authors found that the S elements (red dots) could be seen clearly and are well-distributed within the PA matrix of the 60%-TDC-TFC membrane (TDC to TMC ratio of 6:4) compared to the control membrane (0-TDC-TFC). Besides determining the chemical components, the high-angle annular dark-field STEM (HAADF-STEM) and EELS could provide more opportunities in detecting the segregated solutes at atomic scale.

Applying the same methodology, Goethem et al. [52] investigated the characteristics of nanoparticles (ZIF-8 at concentrations of 0.005 and 0.1 w/v%) embedded within a PA layer. They observed that the HAADF-STEM cross-section image successfully revealed the presence of ZIF-8 particles, although the Zn compounds remained undetected using EDS analysis. This outcome primarily stems from the ChemiSTEM™ detector's sensitivity threshold, which is below 0.03 wt%. Furthermore, the ZIF-8 particle size measured less than 150 nm, featuring an elongated morphology. These findings suggest that the observed particles were not complete ZIF-8 entities, but rather fragments resulting from particle degradation. The authors claimed that Lewis acids such as Zn^{2+} would generate more HCl during the IP process due to the catalytic

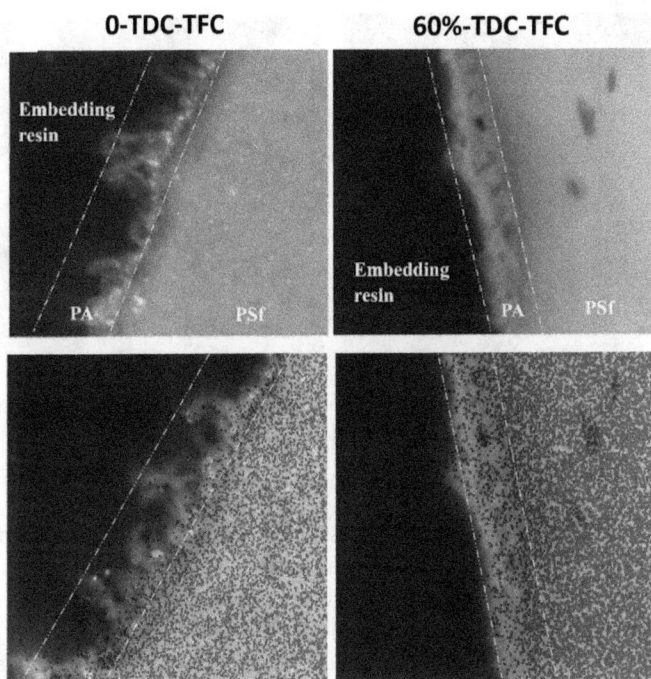

FIGURE 3.11 TEM view (upper) and S (in red dots) elemental map (bottom) of the 0-TDC-TFC and 60%-TDC-TFC membranes under aberration-corrected STEM mode (Note: 0-TDC-TFC and 60%-TDC-TFC represent ratios of TDC to TMC of 0:10 and 6:4, respectively) [51].

effect and thus degrade the ZIF-8 nanoparticles. Currently, the utilization of the HAADF technique for TFC membrane characterization remains quite limited.

Another important feature of STEM is its capability for visualizing the free volume and voids of the PA layer. Employing STEM, Chu et al. [53] investigated the impact of compaction on the reduction of the free volume and thickness of the TFC RO membrane. Figure 3.12 compares the surface area of the PA layer before and after compaction at 60 bar. Obviously, the nodule polymer structure was distorted upon the compaction, which reduced the free volume and thickness of PA layer by 15% and 48%, respectively. Consequently, the water permeability declined from 2.97 to 2.67 L/m².h.bar, attributed to the reduced pathway available for water molecules to traverse.

At its core, STEM and TEM share a fundamental principle, with the key distinction being that the TEM probe does not engage in scanning. In comparison to FESEM, STEM allows for ultra-high-resolution imaging of specific regions during the analysis of the selective layer of a TFC membrane at

FIGURE 3.12   EDS elemental mapping of the cross-section of the TFC membrane (a) before and (b) after being compacted at 60 bar.

Note: green and magenta represent S and N element, respectively [53].

the nanoscale. However, STEM necessitates the use of ultrathin membrane samples, typically less than 100 nm thick. Thus, careful handling is imperative to prevent surface damage and minimize contamination risks, ensuring the production of high-quality scanned images. Another limitation of the STEM technique is its inherently monochromatic image output, which demands a significant degree of interpretation.

3.6 ATOMIC FORCE MICROSCOPY

This section will mainly pivot around the evaluation of emerging atomic force microscope (AFM) as an analysis instrument for TFC surface characterization and fouling. To assess the strengths and limitations of AFM, the consistency, precision and accuracy of results gathered from the reported studies will be reviewed in comparison to the other commonly used techniques. Typically, the measurement of active layer thickness using SEM and TEM is based on visual estimates of the cross-sectional images. This requires subjective judgement to identify the location of the interface between the active layer and support layer, which introduces an additional factor of variability as the two layers are not easily distinguishable [54]. Compared to electron microscopes, AFM can measure the thickness by analyzing the topography profiles of the isolated active layer on a hard solid substrate such as silicon wafer.

Lin et al. [23] evaluated the consistency and accuracy of thickness measurement and surface roughness among commonly used microscopy techniques, namely, SEM, TEM and AFM, and non-microscopy techniques such as ellipsometry and profilometry. Six commercial TFC membranes (i.e., NF270, NF90, XLE, ESPA4, SWC4+ and SW30HR) [23, 55–60], with dry active layer performance levels ranging from NF to RO, were analyzed using each tool to compare the characteristics of the PA layer, to evaluate the versatility of the technique, and to understand the drawbacks of each technique. Specifically for AFM, the tapping mode was used to calculate the difference in the average height of the PA surface and the silicon wafer, while the surface roughness was measured by scanning the AFM tip in the same mode (i.e., tapping mode) to obtain the topography of the membrane surface before determining the root-mean-square roughness.

Table 3.1 summarizes the results, indicating some variabilities in the magnitudes to distinguish the relative thicknesses among the membranes. As an example, the thickness of the PA layer of the NF90 membrane was measured to be an average of about 120 nm via AFM, as shown in Figure 3.13. Using SEM and TEM, the active layers were identified by visual judgement and the PA layer thickness determined as being in the ranges of 184–207 nm and 230–261 nm, respectively. Overall, Table 3.1 shows that (i) SEM and TEM gave significantly larger values than the other methods for all the membranes tested, with the exception of SW30HR; (ii) the values for SEM and TEM were also distinctly different from each another for four out of the six membranes (i.e., NF270 membrane had a thickness of 57 nm as measured by SEM and a thickness of 104 nm when it was measured by TEM) and the error bars are greater possibly due to the subjective judgement required in demarcating the active layer from the support layer; and (iii) the AFM values agree better with those of

TABLE 3.1 Results from Lin et al. [23] for active layer thickness measured for six commercial membranes using different techniques

[a]MEMBRANE	THICKNESS (NM)				
	SEM	TEM	AFM	PROFILOMETRY	ELLIPSOMETRY
NF270	57 ± 2	104 ± 23	21 ± 8	14 ± 5	22 ± 1
NF90	184 ± 23	230 ± 31	120 ± 13	124 ± 8	131 ± 12
XLE	186 ± 27	238 ± 31	136 ± 23	129 ± 11	135 ± 4
ESPA3	155 ± 24	155 ± 23	76 ± 12	88 ± 5	90 ± 3
SWC4+	200 ± 44	340 ± 55	113 ± 14	112 ± 15	116 ± 17
SW30HR	165 ± 22	169 ± 14	176 ± 25	176 ± 23	157 ± 5

[a] NF270 and NF90 are NF membranes; XLE and ESPA3 are brackish water RO membranes; SWC4+ and SW30HR are seawater RO membranes

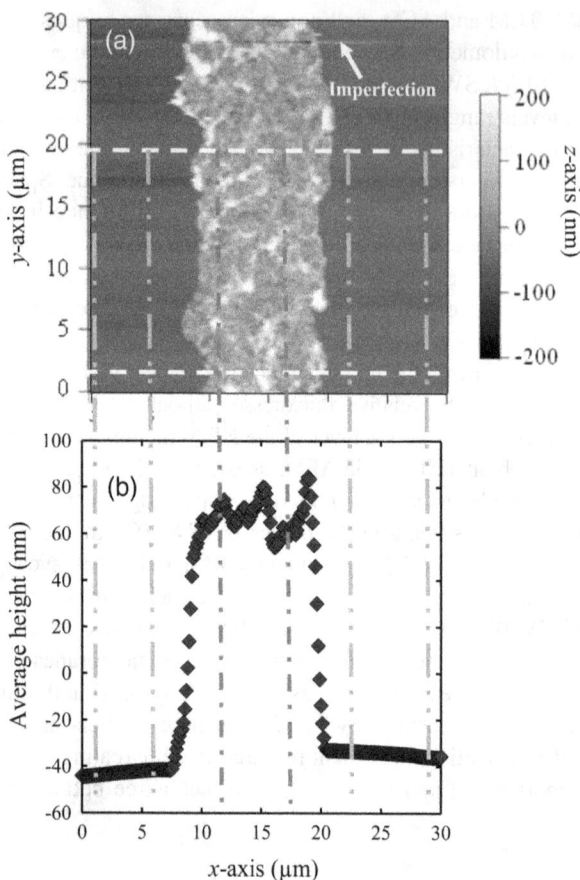

FIGURE 3.13 (a) AFM image of a thin strip of NF90 membrane active layer, isolated on a silicon wafer and (b) corresponding height profile of analyzed region, i.e., between the two horizontal white lines in image (a) [23].

ellipsometry and profilometry for all the membranes tested despite entirely different underlying physical principles [23].

Furthermore, AFM is a useful tool that can be utilized in the development of PA layer to ascertain accurate thickness and surface roughness measurements of PA nanofilms as it can measure down to the nm range in high precision. For the work [23, 61] that explored the effect of controlled IP rate on the formation of sub-10 nm PA nanofilms, AFM was the key tool used to measure the height profile and surface roughness. In the study, all the membranes with varying concentrations had an intermediate nanostrand layer to support the

IP process before it was removed via acid dissolution to form free-standing PA nanofilms and attached to an alumina membrane support. The three main adjustments done were (i) duration of IP between the aqueous and organic monomers to form PA nanofilm, (ii) concentration of the aqueous monomer and (iii) type of aqueous monomer used for IP. AFM was used to compare the height profiles of a fabricated PA nanofilm made 0.1% MPD with an IP duration of 10 min on a silicon wafer to another fabricated PA layer with an IP duration one-tenth of the former (MPD-0.1%-1min). The thickness of the PA nanofilm was 8.4 ± 0.5 nm and 7.5 ± 0.4 nm, respectively, and the root mean square (RMS) roughness was 0.6 ± 0.05 and 0.52 ± 0.04 nm, respectively. With the precise measurements obtained by AFM, the study was able to conclude that a shorter IP reaction time led to a thinner and smoother layer of PA nano-film formed, which correlated to a higher permeance performance.

In addition, AFM was also utilized to study the effect on permeance per-formance when different aqueous monomers were used [61]. Two different aqueous solutions made of 0.1 wt% respectively of PIP and AMP were pre-pared, with all the other parameters and organic phase concentration remaining the same. It was reported that PIP-0.1%-10min and AMP-0.1%-10min had PA layer thicknesses of respectively of 33.2 ± 1.1 nm and 14.5 ± 0.5 nm, and RMS roughness of 4.66 ± 0.25 and 2.31 ± 0.88 nm, respectively. The PA thickness and RMS roughness quantified by AFM correlated well with the organic sol-vent nanofiltration performance, specifically in that the thinner selective layer and lower surface roughness of the PA layer of AMP-0.1%-10min gave higher permeance (Table 3.2). Aside from pure methanol, both membranes were tested with negatively charged dyes with varying molecular weights dissolved in methanol to study solute rejection, with the results shown in Table 3.2. In principle, a thinner active layer increases the concentration gradient across the membrane and a smoother surface contributes to the improvement in perme-ance [62]. AFM is an efficacious tool that can accurately quantify and deter-mine the PA layer thickness and surface roughness for the different fabricated membranes. This enables clear correlation to be drawn between the physical membrane properties and performance.

In another study, Carvalho et al. [63] employed AFM to study the pore sizes of four commercial membranes, namely, two TFC membranes with PA active layers (NF90 and NF) and two asymmetric polyethersulfone (PES) membranes (NP030 and NP010). The 3D AFM surface images showed that a clear difference existed between the two TFC membranes at the nanoscale. The NF90 membrane had larger and mostly circular pores compared to the NF membrane. From the AFM images, the pore sizes of NF90 and NF membranes were quantified to be 6 nm and 4 nm, respectively. Additional exploration demonstrated that the observed permeability of the membranes closely aligned with the calculated values derived from the Hagen-Poiseuille law. The slight

TABLE 3.2 Organic solvent NF performance of nanofilm composite membranes [61]

NANOFILM COMPOSITE MEMBRANE (AQUEOUS-WT%-IP DURATION)	AFM ANALYSIS		PURE METHANOL PERMEANCE (L/M².H.BAR)	NAPHTHALENE BROWN (400.3 G/MOL, 0.95 NM³)		ACID FUCHSIN (585.5 G/MOL, 1.18 NM³)	
	PA LAYER THICKNESS (NM)	RMS (NM)		PERMEANCE (L/M².H.BAR)	REJECTION (%)	PERMEANCE (L/M².H.BAR)	REJECTION (%)
PIP-0.1wt%-10min	33.2	4.66	1.42	1.27	98.2	1.17	99.1
AMP-0.1wt%-10min	14.5	2.31	4.42	4.35	99.8	4.35	99.9

variance in the value is probably attributed to the assumed membrane thickness in the calculations of permeability.

AFM has also been used to investigate the dominant factors associated with the adsorption of foulants on varying nanoscale surface topographies characteristic of membranes. Carvalho et al. [63] studied commercial TFC NF90 and NF membranes before and after fouling by potassium clavulanate (KCA). The fractal dimension and surface roughness were evaluated to correlate with the reduction in flux and fouling performance. The results showed that the NF90 membrane had similar surface roughness before and after the filtration of KCA, indicating only a thin layer of KCA adsorption. On the other hand, the NF membrane surface became smoother after the KCA filtration which was notably different. However, it showed similar granular structures before and after filtration. Furthermore, there was only a minor reduction in the NF permeability performance and KCA solutes have low binding affinity with the PA surface of TFC membranes. Thus, the AFM analysis indicated that the NF membrane had minor accumulation of foulants on the surface. This study evinced that AFM is an effective instrument to gather precise values for fouling effects on the order of nm resolution to draw an accurate relationship between surface morphology and permeability.

Freger et al. [57] modified the TFC membranes by grafting hydrophilic amino acid polymers to mitigate organic fouling. The goal was to have the amino acids densely deposited on the outer layer of the ultrathin PA layer without penetration into the support. ATR-FTIR spectroscopy was used to measure the extent of grafting on the surface by monitoring the changes in absorption peaks. However, the approach was not able to provide adequate details of the grafted polymers. SEM and TEM were not utilized in this work as the PSf background in the SEM images overwhelmed the ultrathin PIP-based active layer, while the TEM images did not provide a clear boundary due to the deep interpenetration between the support layer and PA layer. Thus, the authors employed AFM to elucidate the resulting structures of the various TFC membranes with different degrees of modification. It was proven that clear AFM illustrations of unmodified and varying degrees of modified TFC membranes were obtained, which allowed for quantification of nano-scale differences to relate to fouling extents.

Another aspect that leverages AFM as a tool in membrane studies is the quantification of membrane surface electrical properties for NF membranes in electrolyte solutions of various ionic strength [64]. In the past, surface electrical properties of membranes were commonly evaluated by electrokinetic techniques such as streaming potential measurement. This technique allows the calculation of zeta potential on smooth and homogeneous surfaces, which can be subsequently used in the quantification of interaction between surface and colloidal materials through Derjaguin–Landau–Verwey–Overbeek (DLVO) theory. However, for the materials with surface chemical inhomogeneities and roughness, it only provides calculation of average electrical potential at certain unspecified shear plane. In this context, membranes tend to have significant surface roughness, which makes accurate quantitative evaluation in membrane-colloid interactions in appropriate. Therefore, AFM has been a better choice to measure DLVO-type interactions between a flat surface and a single colloid particle. In such studies, a colloidal particle like silica sphere (2.4–2.75 μm) is immobilized at the tip of the AFM cantilever to directly measure the colloidal interaction and quantify the variation in local electrical double-layer interactions and adhesion on very rough membrane surfaces [65].

With respect to the swelling degree of membrane active layer, Drazevic and Kosutic [60] experimentally investigated the impact of membrane swelling effects on selectivity and permeability using AFM coupled with ATR-FTIR spectroscopy. In this study, the PA layers of four commercial RO membranes (i.e., XLE, ESPA1, BW30 and SWC4+) were isolated on the silicon wafer and gently scratched with a sharp needle to form very narrow strips of PA for AFM imaging in the dry state before it was soaked with deionized water for 30 min to capture the same PA selective layers in the wet state. The average thickness (d) of each PA strip, both in wet and dry state, was quantified as the distance between the two highest peaks in depth histograms. The swelling, Φ, was then calculated from an average of three independent measurements (i.e., three different locations on the narrow film strip) using the following formula [60]

$$\Phi = \frac{\left(d_{AFM,wet} - d_{AFM,dry}\right)}{d_{AFM,wet}} \tag{3.2}$$

Figure 3.14(a-b) shows a flattened image of one of the PA strips for the SWC4+ membrane, and its corresponding height histogram in dry and wet states. The histogram reveals that the SWC4+ PA film has a dry thickness of 122 nm and a wet thickness of 139.3 nm. The average calculated swellings (Φ) for SWC4+, BW30 and ESPA were 0.13, 0.19 and 0.07, respectively. The XLE membrane experienced the highest swelling (Φ) of 0.35, suggesting that the PA layer had the least rigid structure with smaller degree of crosslinking, and gave the highest permeability. Figure 3.14(c) plots intrinsic permeability versus swelling degree for the investigated membranes, along with previous data from other studies. The plot reconfirmed that the obtained selectivity, intrinsic permeability and swelling effects of aromatic PA membranes aligned well with the general trend, affirming the accuracy of AFM in characterizing the intrinsic properties of PA layer.

In 2022, An et al. [66] studied the implementation of morphogenesis concepts to design soft nanomaterials to diversify the functionality of materials.

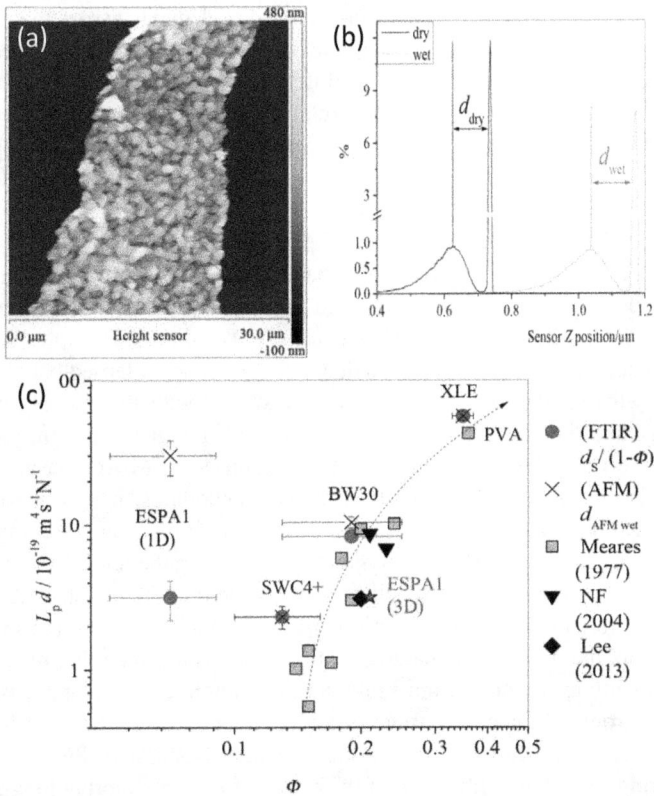

FIGURE 3.14 AFM analysis of one-layer PA of SWC4+ membrane on silicon wafer, (a) flattened topographic image and (b) corresponding histogram. (c) water permeability of different membranes plotted against calculated water content (swelling) [60].

Through the hybridization of liquid-phase AFM and machine learning (ML), they evaluated nanomorphogenesis in the form of 3D crumpling of PA membranes that are often used for molecular separation. The PA layers were synthesized via the IP method to produce active films that are flat but characterizeable by the inner voids of the crumples at the nanoscale. It is crucial to understand the nanomorphogenesis of the crumples as it quantitatively bridges the membrane performance with the fabrication conditions. With the aid of ML, the difficulty of identifying geometric descriptors to represent irregular 3D shapes was overcome, as the developed ML-based procedure was able to identify the critical morphological quantities by converting the surface profiles of solo crumples into 3-nm^2 triangles to collect data on the local Gaussian curvature, curvedness, shape index and fractions of various elements across the surface. Simultaneously, the crumple height, thickness, volume and surface area were extracted from the reconstructed tomographs to determine the eight most important shape descriptors that are related to the local crumple curvature. The eight selected descriptors were then projected into a principal component space to sort the solo crumples into three groups, i.e., dimples, flat pancakes and hollow hemispherical domes [66].

Using the outcomes derived from liquid-phase AFM, An et al. [66] identified a rising pattern in the average apparent modulus. Specifically, dimples, domes, clusters and pancakes exhibited values of 0.56 ± 0.45, 0.57 ± 0.33, 0.90 ± 0.58 and 0.97 ± 0.58 GPa, respectively. The dimple crumples have the smallest apparent modulus which is due to the partially collapsed but extended shapes, whereas pancake crumples mostly sit flat with the largest apparent modulus, and the clusters have approximately 60% higher apparent modulus than dome crumples. With the use of liquid-phase AFM to measure force-indentation curves to determine local apparent modulus, it was proven that mechanical heterogeneity exists in the nanoscale within a single crumple and across crumple groups. Thus, by synthetically altering the relative fractions of the four crumple groups, the mechanical robustness of PA membranes can be controlled to suit the needs of various filtration applications in the industry. In addition, the use of this characterization tool has affirmed that mechanical heterogeneity exists on the nm scale within a single crumple and across the crumple groups. Hence, the usage of AFM as a tool of quantitative imaging has allowed for (i) enhancement of mechanical robustness of PA membranes by altering the relative fractions of the different crumple morphology groups and (ii) understanding the nanomorphogenesis mechanism and relevance to function of soft material systems like PA membranes.

Generally, AFM has multiple advantages compared to other microscopic techniques, including (i) high-resolution images on the order of nm, (ii) direct measurement of thickness without assumption or subjective judgement to distinguish between layers, (iii) quantification of surface roughness and pore size with one prepared sample, as opposed to SEM and TEM analysis which requires two samples, namely, surface to estimate pore size and cross-section to evaluate PA active layer surface roughness [66], (iv) no requirement for sample staining or metal coating as required by SEM and TEM, thus making sample preparation more straightforward [23, 63, 67–69], (v) ability to function in ambient air and liquid environment, making it a useful tool to study biological macromolecules on membranes in comparison to SEM and TEM that requires vacuum environment [70, 71] and (vi) ability to complement with ML studies to explore the nanomorphogenesis of soft nanomaterials like PA which quantitatively links the synthesis conditions with performance metrics such as mechanical robustness and separation efficiency [66].

However, AFM does have limitations as well, including (i) inability to provide details on the chemical composition or distribution of functional groups like FTIR and XPS [69], (ii) small sample size (similar to SEM and TEM analysis), which may not provide a full representation of the actual membrane used for filtration [23], (iii) advantageous high sensitivity can be a drawback particularly for contaminated samples [70] and (iv) white thermal noise may lead to undesirable oscillation of the cantilever tip, which can be filtered by the use of Fast Fourier Transform (FTT) [63, 65].

3.7 SMALL-ANGLE NEUTRON SCATTERING AND POSITRON ANNIHILATION LIFETIME SPECTROSCOPY

Small angle neutron scattering (SANS) and positron annihilation lifetime spectroscopy (PALS) are established non-invasive techniques for material characterization, although their application for TFC membrane characterization is not yet widespread. One of the primary factors contributing to their limited use is the scarcity of these instruments. SANS and PALS have the potential to offer valuable insights into the internal structural characteristics of both the selective and support layers of TFC membranes. Collectively, this information can contribute to a comprehensive understanding of the rejection and permeability performance of a membrane [72, 73].

Scattering of SANS occurs at domains that differ in mass, density and/or chemical composition from the surroundings. It is a quantitative method that penetrates the entire membrane to average over large macroscopic volumes, on the order of 0.1 cm^3, to characterize the pores of the entire TFC membrane [74]. The neutrons scattered by the sample are picked up by the detector and the non-scattered neutrons are absorbed by the beam stop that lies in the center of the detector. The neutrons of different wavelengths are separated by time-of-flight analysis, which enables a wide Q-range and a good Q-resolution simultaneously [74]. The intensity of the neutrons scattered is measured as a function of the modulus scattering vector Q (Eq. (3.3)), and the intensity of q encompasses two contributions, i.e., coherent scattering and incoherent scattering.

$$Q = \left(\frac{4\pi}{\lambda}\right)\sin\left(\frac{\delta}{2}\right)$$

(3.3)

where λ refers to the neutron wavelength and δ is the scattering angle. By deducing the difference between the substrate and the PA-substrate, the scattering of the active PA layer can be determined to obtain the volume fraction and size of PA modules to enable a clearer understanding on the arrangement of monomers on the PA layer [74].

PALS is an analytical technique that employs a positron beam to examine the hole-structure of the functional materials at a molecular level [75] and to quantify the mean size of sub-nm scale holes (i.e., free-volume hole-size) within the polymer chains in thin-film samples [72]. After the positron is injected into a polymer, it is slowed down through inelastic collisions due to ionization and excitation. At the end of its track, the positron annihilates with an electron directly and forms a hydrogen-like atom called positronium (Ps). Ps can exist in two states, i.e., singlet state with antiparallel spins that is known as para-Ps (p-Ps) and triple state with parallel spins called ortho-Ps (o-Ps). The variation in the state is dependent on the relative spin orientation of the electron and positron. In a vacuum-free space, the p-Ps state annihilates into 2γ rays with an extremely short lifetime of 125 ps while the o-Ps state annihilates into 3γ rays with an intrinsic lifetime of 142 ns [72].

In many polymeric materials, o-Ps, a bound state of positron and electron, has a high sensitivity to free volume. The particle is trapped within the selective layer of the free volume until it decays. The lifetime of o-Ps is extracted from the PALS spectra and correlated to the hole dimension, examining the sub-nm sized free volume elements [72]. Aside from measuring the free volume elements' sizes, PALS also quantifies (i) free volume concentration and abundance of free volume elements by determining the relative changes in the intensity (τ_3), (ii) fraction of free volume and (iii) distribution of free volume hole sizes in the range of 1 Å to 10 Å [76]. Furthermore, depth-profiling can be conducted from several nm to several μm of free volume hole radius in the material by altering the incident positron energy (E_{in}) over a wide range of 0.5 keV to 30 keV to study the interfaces, thickness and surfaces of PA active layers [72, 74, 77, 78].

SANS and PALS were previously utilized to evaluate the integrity of the TFC membrane for desalination applications, aiming to comprehend the diverse impacts of operating pressure [79], swelling [76] and morphology of PA layer on the membrane performance [80]. This section will focus on the evaluation of SANS and PALS as a characterization tool to study the PA internal structure and correlate it with the overall membrane performance, including water permeability, solute rejection, compaction effect and fouling. The strengths and limitations of each instrument in comparison to other commonly used tools for PA layer analysis will also be assessed.

In 2020, Pipch et al. [80] demonstrated the benefits of joint exploitation of both SANS and PALS to understand the PA layer and bulk morphology of

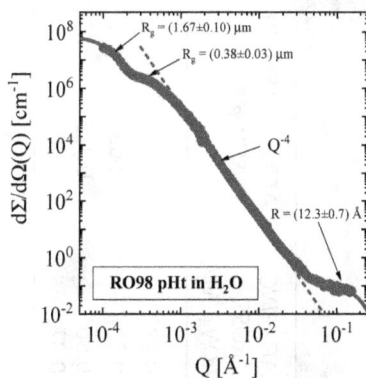

FIGURE 3.15 SANS scattering patterns of the respective layers of RO98 pHt exposed to H_2O [80].

three commercial TFC RO membranes, namely, SW30HR and BW30LE from Dow Filmtec, and RO98 pHt from Alfa Laval. The neutron study based on SANS covering a Q range from 10^{-3} to 0.4 Å^{-1} this allows it to detect pore radii in the range of μm to Å. The scattering patterns of neutrons obtained reflect the morphological details of the entire TFC membrane and identify the various polymers in the respective layers. Coupled with PALS that measures over a depth of approximately 3-μm thickness, the entire PA skin layer was extensively explored to attain structural information of the selective layer.

Figure 3.15 illustrates the SANS scattering pattern of the respective layers, namely, PA, PSf and polypropylene (PP) support layers, of the RO98 pHt membrane. The scattering of the PA layer was visible beyond 5×10^{-2} Å^{-1} from scattering centers and had an estimated gyration radius of approximately 12 Å. It was noted that the scattering from the PA surface layer was too weak to be detected in the Q range of less than 10^{-3} Å^{-1} as the thickness is less than 1 μm, which is close to the detectable limit. However, the scattering range can still be accurately determined by noting the variability of the scattering length density, ρ, between the measured and theoretical values.

Table 3.3 presents the relevant theoretical parameter of the various membrane polymers that are required to analyze the scattering data, while Table 3.4 tabulates the detected scattering length density of the different polymers quantified through SANS experiment conducted by Pipch et al. [80]. The experimental scattering length density data of the SW30HR (2.59 ± 0.03) and BW30LE (2.58 ± 0.03) membranes have excellent agreement with the theoretical value of 2.58 cm^{-2}. In addition, the experimental data of the RO98 pHt membrane (-0.47×10^{-10} ± 0.06 cm^{-2}) agrees well with the theoretical scattering length density of PP (-0.325×10^{-10} cm^{-2}).

TABLE 3.3 Relevant theoretical parameters of the membrane materials [80]

MOLECULE	CHEMICAL FORMULA	MASS DENSITY (G/CM³)	MOLAR WEIGHT (G/MOL)	GLASS TRANSITION TEMPERATURE, T_G (°C)	SCATTERING CROSS SECTION (CM⁻¹)	SCATTERING LENGTH DENSITY, ρ (CM⁻²)
Aromatic PA	$N_2C_4H_{10}O_2$	1.44	238	60–75	0.233	3.10 ± 0.07
Polysulfone	$C_{27}H_{22}O_4S$	1.24	442.54	~220	0.237	2.08
Polypropylene	C_3H_6	0.91	42.08	–10	0.50	–0.325
Polyester	$C_{10}H_8O_4$	1.38	192.17	~79	0.37	2.58

TABLE 3.4   SANS experimental parameters of membrane materials [80]

MEMBRANE	POLYMER	SCATTERING LENGTH DENSITY, ρ $(\times 10^{10}$ $CM^{-2})$	GYRATION RADIUS (PORE SIZE) (μM)
Alfa Laval RO98 pHt	PA-polysulfone-polypropylene	$-(1.11 \pm 0.11)$	1.10 ± 0.01
Alfa Laval RO98 pHt	PA-polysulfone-polypropylene	$-(0.47 \pm 0.06)$	1.30 ± 0.01
Dow Filmtec SW30HR	PA-polysulfone-polyester	2.59 ± 0.03	1.25 ± 0.01
Dow Filmtec BW30LE	PA-polysulfone-polyester	2.58 ± 0.03	1.26 ± 0.02

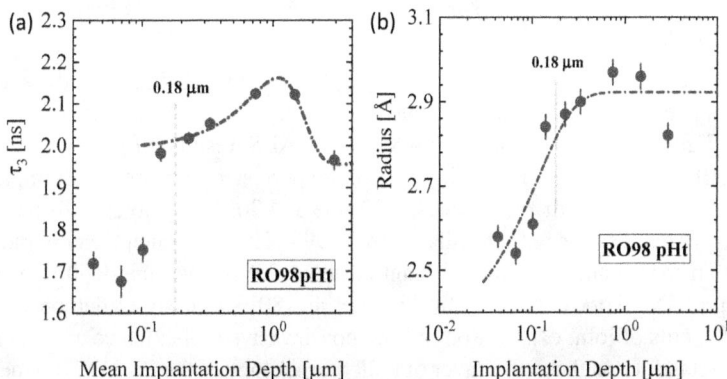

FIGURE 3.16   (a) Lifetime of RO98 pHt membrane against mean implantation depth and (b) Pore radius in the RO98 pHt membrane against implantation as determined from the lifetime spectra [80].

PALS was also utilized to cross-analyze the characteristics of the commercial TFC RO membranes. By using the slow positron method, the implantation depth of PALS can be adjusted to determine the nm-sized pores at distances up to 4 μm from the membrane surface. Figure 3.16(a) and (b) shows the lifetime (τ_3) and radius of RO98 pHt membrane with varying implantation depths, respectively. The green dashed/full lines in the images denoted the interface between the PA and substrate layers. Using the lifetime trend from the figure, the radius of the RO98 pHt membrane pores could be estimated. At a 0.02-μm surface layer depth, the pore radius was 2.5 Å and increased consistently to about 2.9 Å as the surface depth increased to 0.18 μm. Subsequently, the pore

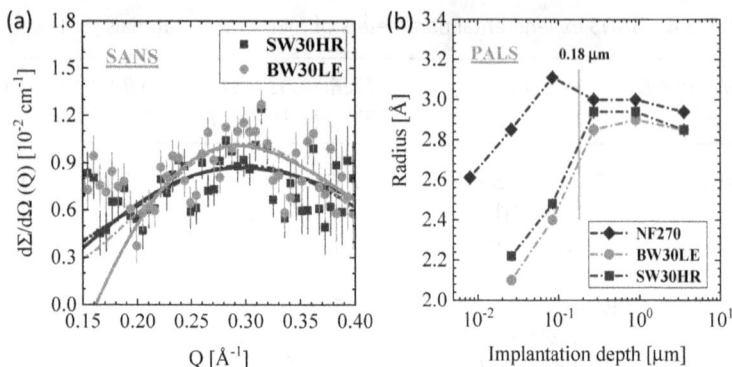

FIGURE 3.17 (a) Large Q SANS data in vacuum from the same membranes and (b) Radius of pores against implantation depths for different RO and NF membranes with PALS [80].

radius remained relatively constant at 2.9 Å, which represents both the selective active layer and porous substrate.

Figure 3.17 presents the SANS and PALS results of BW30LE and SW30HR membranes. Regarding PALS, the pore size increased with implantation depth for both membranes, with the BW30LE membrane displaying slightly smaller pore radii. Similar to the RO98 pHt membrane, the pore radius of both membranes became constant at about 2.9 Å once the depth reached 0.2 μm. The work conducted by Pipch et al. [80] successfully demonstrated the benefits of joint exploration of both non-invasive tools to have an in-depth understanding on the active layer of bulk morphology of the PA TFC RO membranes with different properties.

Another SANS study was conducted on similar standalone aromatic PA films to examine the isolated active layer in dry and wet conditions [74]. Twenty PA films were prepared with the same IP technique and separated into two stacks of 10 PA layers each (denoted as PA-1 and PA-2, respectively) by random prior to analysis to ensure adequate scattering detection by the neutron beam on the internal PA pores. With reference to Figure 3.18, the PA-1 stack gave a radius of 7.8 ± 1.2 Å while PA-2 stack gave a radius of 8.3 ± 0.6 Å. This low variability of PA radius implies that SANS has a relatively good precision in obtaining intricate internal details. Aside from quantifying the pore radius, the average volume fraction and scattering contrast were also obtained. The results showed that the average volume fraction for PA-1 and PA-2 was about 28% and 21%, respectively, indicating that nearly 30% of the surface of PA globules was covered with small nanoscopic pores that give rise to the fractal structure. The additional details of surface characteristics allow thorough

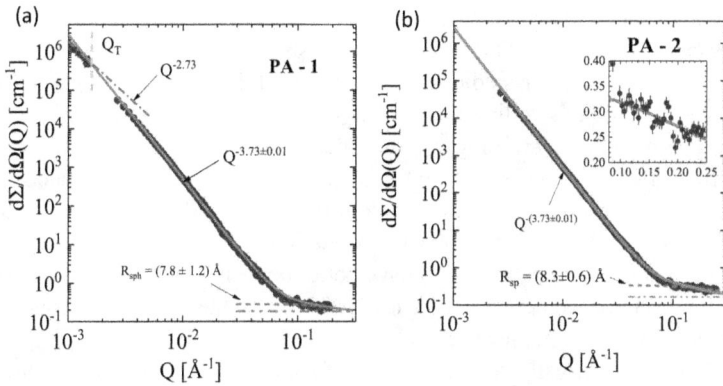

FIGURE 3.18 SANS scattering pattern for (a) PA-1 stack and (b) PA-2 stack. Each stack consists of 10 PA active layers [74].

evaluation of the PA surface layer, which allows for correlating membrane performance and active layer properties.

SANS is also able to accurately measure the pore size of PA layers in dry and wet environments. Figure 3.19 illustrates the scattering data of PA-2 stack foils before and after it was exposed to D_2O solution for 12 h to simulate dry and wet PA layers [74]. The pore sizes of the dry and wet PA membranes were also similar at 1.41 ± 0.02 μm and 1.39 ± 0.02 μm, respectively. When compared with molecular dynamic (MD) simulations, the measured pore volume fraction of the hydrated PA film was about 30%, similar to the theoretical water uptake of 27% in the PA. Thus, despite the weak scattering of neutrons on the PA surface layer in a typical TFC membrane, the data obtained from SANS are

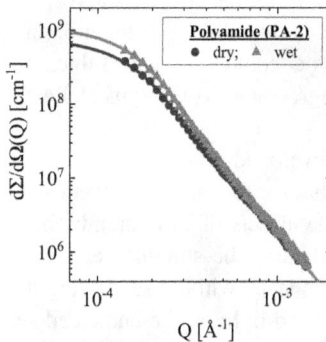

FIGURE 3.19 Scattering data from dry and wet PA-2 stack foils after exposition to D_2O for 12 h [74].

fairly accurate. Complementing SANS results with that from other tools will enhance the quality of TFC membrane analysis.

Furthermore, other studies reported the usefulness of utilizing SANS to confirm the unique swelling mechanism of TFC membranes for desalination [81]. The membrane swelling was revealed to be relatively uniform, but the swelling effect was not significant enough to change the length scale in the PA structure. These results were also confirmed by SANS measurements done by Pipch et al. [74], and further works were highlighted to reduce the tortuosity of the convective pathways to have better control of the polymer network structure to improve permeance. Thus, SANS was able to valuably contribute to improve PA TFC permeance.

Aiming to investigate the effect of free volume characteristics on salt rejection and water permeation of RO membranes, Li et al. [78] varied the positron incident energy of the F1 (1 wt.% DMSO) TFC RO membrane. They found that, by increasing the incident energy from 0.5 to 1.0 keV, the o-Ps lifetime increased rapidly from 1.36 to 1.82 ns before approaching the maximum of 1.87 ns at around 2.0 keV, at which most of the positrons were implanted in the selective PA layer. Results also showed that the o-Ps lifetime decreased slightly when the positron energy increased to 3.0 keV. This may be attributed to the thin dense back side of the PA layer that possessed a small free volume. With further positron energy increment to 10 keV, the majority of the positrons passed through the active layer and were embedded in the PSf support where it was annihilated. Thus, the o-Ps lifetime reached an observable gentler gradient before plateauing. The long lifetime, o-Ps, is well-correlated with the mean free volume hole size as described by the Tao-Eldrup equation. The variation of o-Ps intensity as a function of the implantation depth was well-fitted to identify the sub-surface of the PA layer, active layer and porous support. The fitted o-Ps intensities of the sub-surface-active layer, PA selective layer and support were quantified to be 5.1%, 12.1% and 22.6%, respectively. Hence, the fitting of the experimental data enabled the quantification of thickness across various layers. The PA layer was concluded to have a thickness of approximately 200 nm, which is in close agreement to the typical PA active layers of TFC RO membranes [75].

Zheng et al. [82] developed a novel ionic liquid (IL) assisted method to fabricate PA TFC membranes for organic solvent nanofiltration (OSN) applications. They studied the effects of TFC membrane when monomer concentration was altered to identify the suitable selection of mixed monomers to boost the membrane permeance without sacrificing the rejection performance. PALS was then used to shed light on the enhanced separation and permeance performance, and the potential of recyclability of ILs for sustainability when harsh organic solvents like n-hexane was replaced with ILs in the PA TFC membrane fabrication process. Through the results gathered by PALS and other characterization tools, understanding was obtained on the mechanism of ILs that aid in decreasing the diffusivity of monomers in the organic phase, which decreased the cross-linking degree of the PA layer and thereby reducing the reaction rate. Figure 3.20 illustrates the S parameter changes of the three fabricated membranes, i.e., PMIA/PA1-IL (PEI and TMC in ([BMIm][Tf$_2$N])), PMIA/PA4-IL (PEI/PIP and TMC/PDC in [BMIm][Tf$_2$N]) and PMIA/PA4-HX (PEI/PIP and TMC/PDC in n-hexane), as a function of positron incidence energy and the corresponding penetration depth. At the peak, the PMIA/PA4-HX membrane had the smallest S-parameter, which correlates to the smallest free volume, indicating it as the densest membrane with lowest

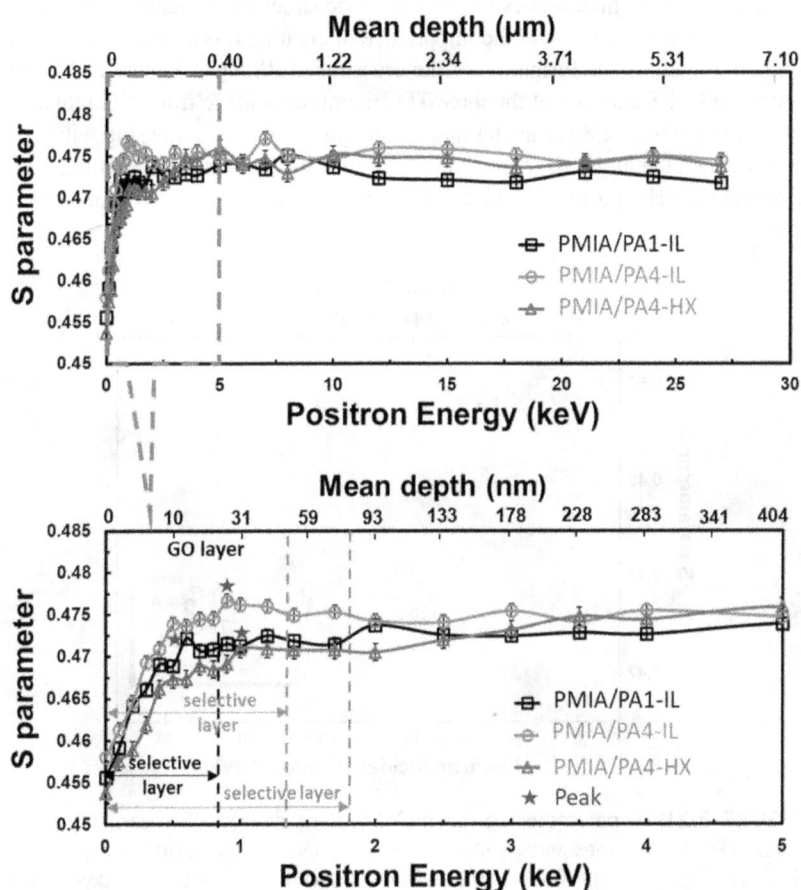

FIGURE 3.20 S parameter curves of the fabricated PMIA/PA1-IL, PMIA/PA4-IL and PMIA/PA4-HX membranes as a function of positron energy [82].

permeance. On the other hand, the PMIA/PA4-IL membrane was the least dense membrane as it has the greatest free volume content. The thickness of the selective layer was consistent with that obtained from SEM imaging. By combining PALS with other characterization tools, it can be affirmed that the incorporation of IL can reduce the diffusivity of monomer in the organic phase, which reduces the crosslinking degree of the PA layer as the reaction rate decreases. This enhances the membrane performance and recyclability, while ensuring it has sufficient mechanical and tensile strength for practical usage.

In 2022, Lee et al. [83] employed PALS to analyze the free volume of the PA TFC membranes to study the effects of monomer rigidity on microstructures for pervaporation dehydration applications. Different diamines were used to fabricate the TFC membranes via IP. The steric structure of diamine molecules was evaluated to understand the importance of creating less compact structures without impeding the formation of the crosslinked PA layer. Figure 3.21 illustrates the PALS analysis of the three TFC membranes made from different aromatic diamines. The S parameter has a direct proportional relationship with free volume, i.e., the larger the S parameter value is, the greater the free volume. In a region whereby positron incident energy was less than 1.5 keV (purple dotted

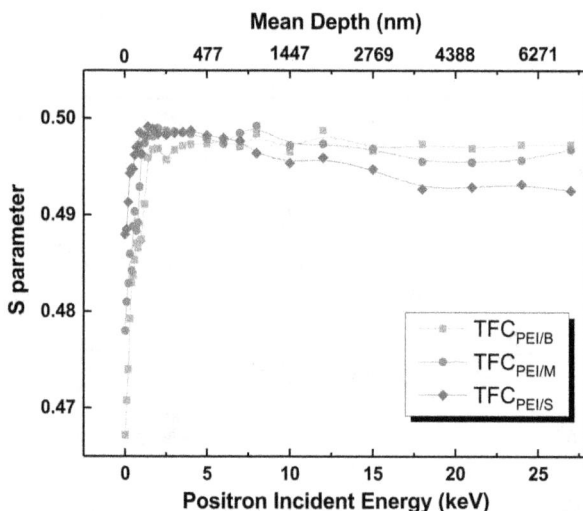

FIGURE 3.21 S parameter VS positron incident energy VS mean depth for three TFC membranes with varying diamines. (Note: The monomer used for TFC$_{PEI/B}$, TFC$_{PEI/M}$ and TFC$_{PEI/S}$ membranes was 4,4′-((propane-2,2-diylbis(4,1-phenylene))bis(oxy))dianiline (BAPP), 4,4′-((methylene bis(4,1-phenylene))bis(oxy)) dianiline (MPDA) and 6,6′-bis(4-aminopheoxy)-4,4,4′,4′,7,7′-hexamethyl-2,2′-spirobichroman (SBC), respectively) [83].

frame in Figure 3.21), the S parameter rapidly increased. This is due to the diffusion and scattering phenomena when the positrons are entering the polymer surface. The TFC$_{PEI/B}$ membrane had the smallest S parameter value in the PA layer, which indicated that the monomer had the lowest free volume, and produced a denser selective layer compared to the other two monomers. On the other hand, the TFC$_{PEI/S}$ membrane showed the highest S parameter value, which indicated the greatest steric hindrance due to the rigid and twisted spirobichroman structure that reduced chain packing. Through the usage of PALS, monomer rigidity on microstructure can also be extensively studied to find a suitable monomer to produce a defect-free TFC membrane for a range of applications that may require different compaction levels. Based on the PALS characterization done in this work, it was concluded that the TFC$_{PEI/B}$ membrane's rigidity led to enhanced overall membrane performance and property that resulted in longer lifespan at comparable permeation flux with other reported membranes.

Furthermore, PALS is capable of analyzing the inner-layer properties of TFC RO membranes after subjecting to a harmful compaction process. Aiming to evaluating the effects of high applied pressure compaction on the physiochemical properties and permeability of RO membranes, Chu et al. [53] prepared six identical TFC membranes and compacted them at different pressures ranging from 10 to 60 bar. The thicknesses and free volumes of the PA active layers were then quantified with PALS. Table 3.5 compares the relevant PALS results of the TFC membranes with and without compaction. Based on the o-Ps parameters, the free volume radius and free volume can be directly calculated to compare the changes in the inner structure after the compaction. The higher the

TABLE 3.5   Free volume size and intensity, and thickness of active layer and TFC RO membranes measured by PALS (positron incident energy = 1 keV, compaction duration = 16 h, density of PA layer = 1.3 g/cm³) [53]

COMPACTION PRESSURE (BAR)	O-PS LIFETIME T_3 (NS)	I_3 (%)	FREE VOLUME RADIUS (NM)	FREE VOLUME ($Å^3$)	TOTAL THICKNESS (μM)	PA ACTIVE LAYER THICKNESS (NM)
Pristine	1.63	9.31	0.263	69.456	258.2	425.2
10	1.61	9.26	0.255	63.892	156.5	392.7
20	1.59	9.15	0.244	60.850	151.2	326.0
30	1.58	9.14	0.242	59.456	144.5	302.3
40	1.56	9.13	0.235	59.366	140.8	296.9
50	1.60	9.18	0.233	52.985	140.3	232.3
60	1.58	9.15	0.224	47.080	124.7	221.2

compaction pressure, the smaller the o-Ps lifetime is, which results in a smaller free volume radius and thinner active layer. The original PA active layer thickness was significantly reduced by approximately 48% when 60 bar was applied on the TFC RO membrane. This reduction caused a negative impact on membrane permeance. It was also reported that the membrane permeance performance was primarily affected by the amount of PA layer compaction as the nodular polymer structures were distorted and the change in viscoelastic polymer properties were more significant in the PA selective layer than the support layer.

Like all analytical equipment, SANS and PALS have specific strengths and limitations. Both techniques are non-invasive and do not require special sample pre-treatment prior to sample analysis. Furthermore, both tools have great versatility and flexibility, allowing them to be used to evaluate different sample types in various conditions, such as in wet state [72, 74, 84, 85] and under atmospheric conditions within a humidity range [72, 85, 86]. Thus, the PA layer does not have to be altered before it is analyzed, providing the best representation of the PA layer in the natural state.

However, it must be pointed out that SANS is a labor-intensive approach that requires some rectification before it can be accurately analyzed. This is mainly due to poor localization of strong scattering. Thus, corrections for multiple scattering and performing contrast variation measurements must be done to resolve and enable identification for more detailed morphological characterization of the membrane [80]. Another drawback of SANS is its limitation to isolate the PA layer from the substrate layer. The scattering of neutrons from the active skin is too weak to be distinguished from the total scattering [74, 80, 87]. Thus, many studies couple PALS with SANS for membrane analysis to thoroughly quantify the PA layer [80]. Lastly, studies have observed that the statistical error of SANS data tends to be marginally larger compared to PALS [74].

PALS is limited with respect to measuring free volume fraction and free volume shape. Thus, a better understanding of solute transport can only be possible through the combination of free volume hole size derived from PALS and MD simulations at this point. In addition, the quantification of free volume via PALS, which is fitted into the Tao-Eldrup model, is restrictive with regard to the assumptions that all holes are spherical. This assumption might not hold for the cross-linked PA layer.

SANS and PALS are useful tools that complement each other in the investigation of the inner structure of the PA layer, which is the critical layer in a TFC membrane as it directly impacts the solute rejection and flux performance. By attaining comprehensive details on the internal structure of the active layer, TFC membranes can be better understood and optimized to address the current challenges in various industries. Aside from collating extensive knowledge of the PA layer, it is also important to study the two support layers as it may have an indirect impact on the membrane performance over time due to compaction or swelling effects [79].

3.8 MEMBRANE STABILITY TEST BASED ON T-PEEL METHOD

To evaluate the adhesive properties and structural stability between the substrate and the PA selective layer of TFC membranes, some researchers employed the T-peel method to analyze the interaction among the layers [88]. The letter "T" originates from the shape created by the substrate and the polymer layer being simultaneously removed. This is achieved by holding one end tightly in the jaw while the other end is pulled off in a vertical direction as illustrated in Figure 3.22(a) [89]. This technique is similarly employed for TFC membranes, where the film is joined by either rolling or bonding with adhesive tape, creating a loading grip. The length of the cling area is approximately 10 mm, while the sample itself measures 10 mm in width and 25 mm in length [90]. It should

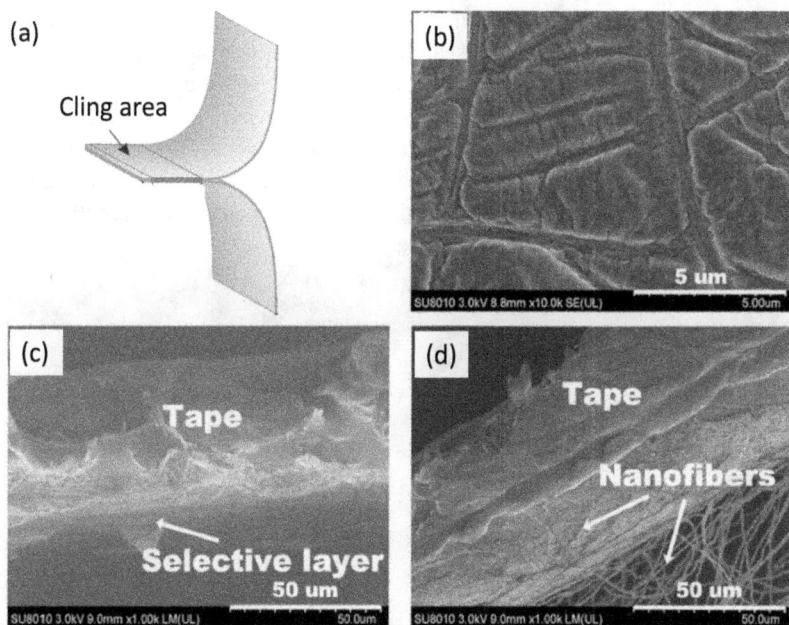

FIGURE 3.22 (a) Schematic of T-peel test configuration showing its loading and cling area [89] and SEM images of TFC membranes after T-peel strength test, (b) back surface of control membrane (TFC-0) after the T-peel strength test, (c) cross-section of control membrane (TFC-0) and (d) cross-section of modified TFC membrane (TFC-12) after the T test. Note: The TFC-12 membrane was fabricated using PDA-coated nanofiber and the amount of PDA in the coating solution is 12 mL [90].

be noted that the peel strength is always recorded as the maximum peel force between both layers and the results are obtained with the help of a computer-controlled universal testing machine.

Luo et al. [90] fabricated TFC membrane by forming cross-linked PA layer on the surface of polydopamine (PDA)-coated polyacrylonitrile (PAN) nano-fiber mat. When contrasting the T-peel test outcomes of the control membrane (without PDA coating) with those of the TFC membranes produced using a PDA-coated substrate, a notably elevated adhesion strength was observed in the modified membranes. Specifically, the adhesion strength between the PA layer and the substrate rose substantially from 0.1 N to 0.29 N. As shown in Figure 3.22(b–c), the selective layer of the TFC-0 membrane (control) was peeled out of the substrate completely, leaving only some imprints of the nano-fibers on the back surface after the T test. However, for the TFC-12 membrane, its selective layer was found to adhere firmly on the substrate and damage on the substrate surface could be clearly from the cross-section image (Figure 3.22(d)). This phenomenon can be attributed to the chemical bonding and interlocking between the PA and PDA, as well as the robust adhesion existing between PDA and the nanofiber mat.

Presently, there is limited research effort that utilize the T-peel method to evaluate the interplay between the PA layer and the substrate. The most challenging aspect of this technique revolves around the initial removal of the PA thin film from the substrate. Throughout the analysis process (i.e., during the peeling off), it is imperative to take measures to prevent any harm or fracture to the PA film, ensuring its integrity is maintained and the adhesive strength could be measured by tensile testing machine.

Filtration-Based Techniques/ Approaches for TFC Membrane Characterization

4

4.1 FILTRATION USING GOLD NANOPARTICLES

The PA layer in a TFC membrane typically exhibits an irregular and compact structure, often presenting uneven ridge-and-valley patterns or nodular formations. Examination and measurement of this structure are commonly conducted using electron microscopy techniques. However, microscopic examinations do not provide a comprehensive understanding of the inherent attributes of the PA layer. Thus, Tan et al. [91] conducted membrane filtration tests by filtrating gold nanoparticles (GNPs) with a concentration of 1.0×10^{12} ppm at 4.8 bar for 10 min. Subsequently, they employed microscopic characterization to visually inspect the spatial distribution of water permeability sites within the Turing-structured (TS) membranes. It was revealed from TEM micrographs that the deposition of GNPs was not uniformly distributed on the PA surface. Two TS membranes were investigated in this study, with TS-1 and TS-II denoting membranes with nanoscale spots and stripes, respectively. The TS-II membrane exhibited higher water flux of 125 $L/m^2.h$, which was approximately two times higher than that of the TS-I membrane of 64 $L/m^2.h$, while maintaining similar Na_2SO_4 rejections (>99%) under the same testing conditions (measured at 4.8 bar, 2000 ppm Na_2SO_4). This finding

DOI: 10.1201/9781032690346-4

exhibited a strong correlation with the surplus surface area ratio of the membranes. This correspondence suggests that the Turing structures exert a significant influence on the water flux. Thus, it was concluded that certain distinct regions within the Turing structures gave relatively heightened water permeability, which contribute to the membranes exhibiting improved water transport characteristics.

With the aim of investigating the flow activity of PA layer in a commercial TFC RO membrane (SW30RH FilmTech™), Li et al. [92] infiltrated different GNP solutions (containing 1-, 5-, 10-, 20- or 40-nm GNP at 5×10^{12} ppm) through the membrane at 3 bar. The utilization of various nanoparticle sizes serves the purpose of assessing the fluid dynamics of the nodular surface of the PA and estimating the dimensions of the pathways for permeation. To gain insights into the connectivity of the void regions on the PA surface, back-filtration experiments using 5-nm GNP were initially conducted. TEM image results as shown in Figure 4.1(a,i) revealed that the 5-nm GNP seemed to reach

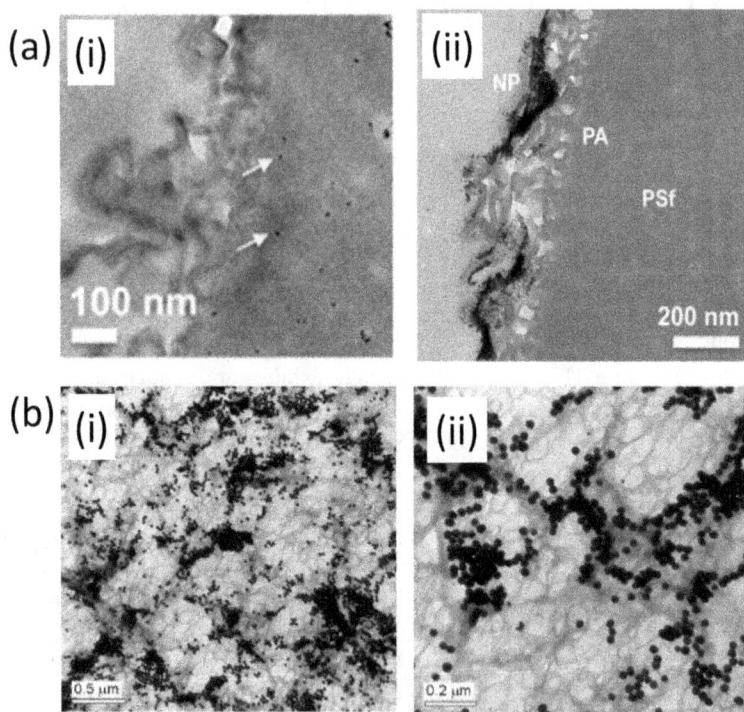

FIGURE 4.1   (a) TEM cross-sectional images of RO membranes with (i) 5-nm GNPs back filtered into the membrane and (ii) 5-nm GNPs front filtered onto membrane active surface [92]. (b) TEM images of commercial membrane (ESPA3) containing 30-nm GNPs deposited during a dead-end filtration experiment at scale bar of (i) 0.5 μm and (ii) 0.2 μm [93].

the interface between the PA layer and the PSf layer, positioned just beneath the rear surface of the PA. However, when larger GNPs (10, 20 and 40 nm) were used, the nanoparticles were completely retained at the PSf layer as they were able to penetrate the larger voids in the PSf. Using 5-nm GNP solution but under the normal mode (i.e., front filtration), the TEM image (Figure 4.1(a,ii)) showed that the top surface of the PA layer was covered with nanoparticles. These particles agglomerated predominantly, forming a valley-like region on the selective layer surface. The observation indicated that a portion of the GNPs managed to access void-like regions situated in the upper segment of the PA layer. This observation serves as confirmation that these void-like regions possess direct connectivity to the outer surface. In other words, this observation demonstrated that the pore size of the PA layer was smaller than 5 nm, rendering it effective in retaining the 5-nm GNPs.

Similar to Li et al. [92], Pacheco et al. [93] studied the PA structure of a commercial TFC membrane (ESPA3, Hydranautics) through GNPs filtration method operated at 5 bar. Differing from other studies that infiltrated GNPs of varying sizes through the membrane, this study employed GNPs with a consistent size of 30 nm. These 30-nm GNPs were applied to the membrane's surface and subsequently visualized using TEM. As shown in Figure 4.1(b), particles were well distributed over the entire thin film. GNPs were found to accumulate over the dark regions (ridge structure) rather than the bright regions (valley structure) of the PA layer, indicating ridge structure is the loose part of the PA layer.

Analyzing the PA layer's structure through the utilization of the GNPs filtration technique, followed by projected area TEM, presents a novel approach for visually characterizing the isolated PA layer. This method eliminates the need for elaborate sample preparations such as resin embedding for cross-section TEM or applying metal coatings for SEM. Nonetheless, the principal concern for researchers remains the substantial cost associated with conducting GNPs filtration experiments.

4.2 IN SITU/ONLINE FOULING CHARACTERIZATION

4.2.1 Optical Coherence Tomography

Optical coherence tomography (OCT) is an advanced optical imaging technology that has gained prominence in the field of membranes in recent times. This non-invasive technique offers the capability to assess membrane fouling

[94, 95]. Making use of its unique characteristic of low-coherence interference, OCT can generate depth profiles of semi-transparent samples. It achieves this by utilizing near-infrared light with wavelengths ranging from around 900 to 1300 nm. These depth profiles can then be combined to create a 3D image, giving a depth resolution of approximately 2 μm. Compared to other tomography techniques such as X-ray computed tomography, OCT is able to perform the cross-sectional imaging at a relatively fast rate of tens of kilohertz [96, 97].

Sample visualization via OCT is based on the optical interference between the reference light and the light reflected from the sample. The light source is usually a super luminescent diode that can emit near-infrared light. By incorporating the Doppler effect, OCT characterization goes beyond merely capturing structural images and extends to capturing velocity profiles as well. This enhancement enables the extraction of not only visual information about the structure but also insights into the velocity patterns within the examined samples.

Quantitative analysis of the growth of the fouling layer during membrane-based filtration has been achieved by analyzing a series of real-time OCT scans with image processing techniques based on background subtraction [98, 99]. An attractive function of OCT is not only the Doppler imaging that visualizes the velocity profiles of a fluid field, but also the ability to monitor in real-time the evolution of internal fouling within the membrane [100].

One of the pioneering studies concerning the application of OCT for membrane characterization was conducted by Gao et al. [101]. In this study, an OCT system was integrated with a laboratory-scale membrane filtration system to observe the growth of the fouling layer (caused by bentonite microparticles) on a membrane during filtration. Figure 4.2(a) presents the spatial domain under examination and the specific cross-section that was scanned using OCT. This cross-section is denoted by the red color and was captured after introducing the detection light beam into the filtration module's channel. The crossflow velocity during this process was maintained at 3 cm/s. The direction of the bulk flow was parallel to the cross-section and a Doppler image was first obtained to verify the flow pattern in the channel. The dashed-dotted curve was added into the OCT images to help distinguish the membrane surface. While the membrane was being fouled during filtration, a series of structural images were continuously recorded in the primary cross-section. Figure 4.2(b)–(c) shows two structural images captured at 0 min and 60 min, respectively. Compared with the clean membrane at 0 min, the profile of the fouling layer could be easily identified after 60-min filtration. An additional OCT scan in the flow direction perpendicular to the cross-section was also implemented at the end of the filtration and the result is displayed in Figure 4.2(d). Both the axial and transverse profiles indicated that a uniform cake layer was deposited on the membrane surface after filtration. It should be noted that there are some discernible changes of the gray scale from the center of the channel to both

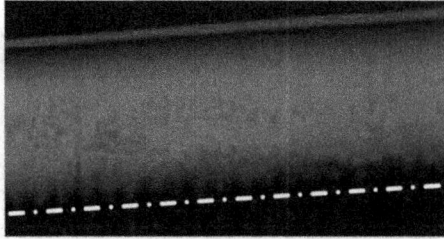

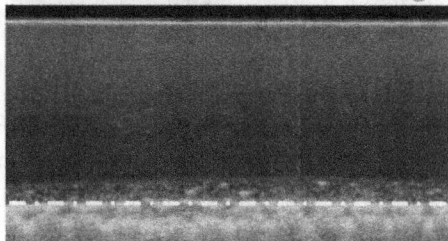

FIGURE 4.2 Primary OCT cross-section images of the fouling including: (a) the Doppler image of the velocity field in the channel at t = 0 min, (b,c) structural image at t = 0 min and 60 min and (d) the structural image in the channel at t = 60 min. The secondary cross-section scanned by the OCT was perpendicular to the direction of the bulk flow as indicated by the red area in the lower left schematic.

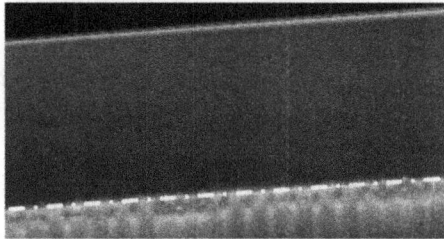

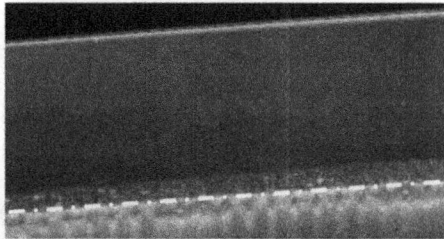

Note: The direction of the bulk flow is denoted by the arrow while the membrane surface is indicated by the dashed–dotted curve [101].

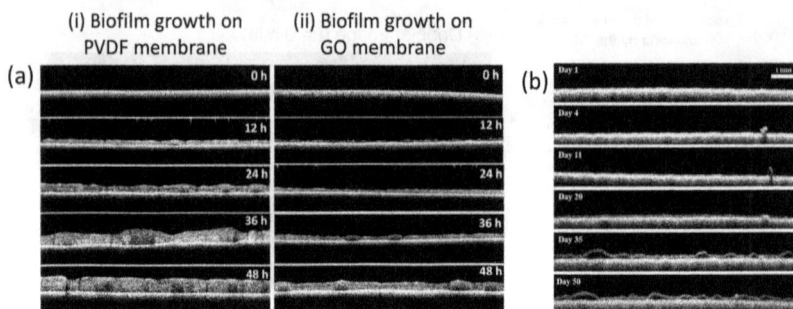

FIGURE 4.3   (a) OCT images of biofilm growth behavior on the (i) PVDF and (ii) PVDF-GO RO membrane during continuous flow operation for a period of 48 h [102]. (b) Evolution of a 2D biofilm on a membrane over the course of 50 days at a fixed location [97].

membrane surface and upper channel wall, i.e., the observation window. The variation of the gray scale suggested the change of the fluid's scattering characteristic caused by the variation of the concentration of the foulants.

Using OCT, Farid et al. [102] evaluated the biofilm development and antibacterial adhesion of a GO-coated RO membrane during filtration of gram-negative *E. coli* in dead-end mode and compared the performance of this membrane with the control membrane (without GO) at different time intervals. The coated membrane was fabricated by filtering GO suspension through PVDF membrane via a vacuum filtration approach. The OCT images as shown in Figure 4.3(a) indicated the biofilm development on the virgin PVDF and PVDF-GO membranes over a period of 48 h. At the early stage of experiment ($t = 0$ h), the OCT image showed a clear surface for both types of membranes. The biofilm initiation began at $t = 12$ h and continued to grow as filtration progressed. Gradually, this led to the formation of a cake-like layer structure with a notably thickened morphology. Nevertheless, it must be pointed out that the thickness of the biofilm on the modified RO membrane surface was significantly lower than the control membrane after the 24-h operation. Following a 48-h operation, the biofilm thickness on the modified RO membrane was measured at 26.1 μm and this was considerably less than the biofilm thickness observed on the control RO membrane (56.5 μm). The observation clearly suggested the positive role of GO in enhancing the anti-biofilm/biofouling activity of the RO membrane.

Park et al. [97] employed OCT to investigate biofilm development on a RO membrane. They subjected the membrane to an extended filtration procedure lasting up to 50 days using feed water containing *Pseudomonas aeruginosa* bacteria (PAO1). As shown in Figure 4.3(b), the membrane remained visibly clean for the initial 4 days. However, a discernible biofilm became apparent on the membrane surface by day 11 of the experiment. From day 20 onward, a bright-orange coloration could be seen on the membrane surface. From day 35 to 50, an obvious change in membrane morphology was detected, owing to the growth rate of the biofilm being nearly equal to the detachment rate caused by the cross flow on the biofilm structure. From the OCT images, it was observed that surface conditioning did not take place until day 11 of the experiment as there was no visible change on the membrane surface caused by the deposition of organic matters. The results indicated that the bacteria growth rate was the highest between day 20 and day 35 of the experiment. After day 35, there was minimal alteration in the thickness of the fouling layers. This implied that, beyond day 35, the growth rate of the biofilm was approximately equivalent to the rate at which it detached from the membrane surface.

Although OCT offers a more convenient and efficient approach for diagnosing membrane fouling, its application is hindered by the inherent trade-off between axial resolution and imaging depth (typically around 2–3 mm). This limitation constitutes a significant challenge to the broader implementation of OCT in membrane studies. Additionally, the complexity of image analysis and high cost of OCT equipment are other notable drawbacks that need to be considered when utilizing OCT for membrane characterization.

4.2.2 Electrochemical Impedance Spectroscopy

Another non-invasive method for characterizing membranes is electrical impedance spectroscopy (EIS), which has been employed to investigate the electrical characteristics of TFC membranes during fouling and cleaning procedures [103, 104]. EIS is capable of detecting changes in the structural layers of the membranes and can distinguish the individual conductance and capacitances of both the PA layer and the substrate. The underlying principle of EIS involves introducing sinusoidal alternating currents at various predefined frequencies into a system [105]. The current and the corresponding voltage across the sample, including the phase difference between the voltage and current, are used to deduce the capacitance and conductance of the system at each frequency [106, 107]. The variation of the capacitance and conductance can then be used to determine the number and properties of the layers/elements of the system, enabling membrane fouling to be assessed during filtration [108].

Sim et al. [106] quantitatively described the variation of the electrical properties of a commercial RO membrane (BW30, Dow FilmTec™) using EIS as fouling developed. The internal structure of the membrane was represented by a series of parallel conductance and capacitance elements. Upon the introduction of a mixture containing alginate (10 ppm) and silica (2000 ppm) at a pressure of 0.96 bar, a reduction in the conductance of all electrical components was noted, decreasing from the initial value of 15.9 S/m^2. This reduction was prominent at 4.8 h, when the conductance dropped to 12.7 S/m^2. However, as the fouling continued, the conductance exhibited an upward trend, rising to 13.7 S/m^2. This is because, at the initial filtration stage, the non-conductive alginate and silica that deposited on the membrane surface decreased the overall conductance of the skin layer. However, when the foulants started to build up as filtration progressed, the cake tended to enhance the concentration polarization effect due to the increased NaCl concentration near the membrane surface. Figure 4.4(a) shows the changes in the Nyquist plots during the fouling at different times. In order to establish the reversibility of the foulant layer, the researchers subjected the fouled membrane to flushing with a 2000-ppm NaCl solution for approximately 17 h. It was observed that the membrane reverted to its initial condition, with capacitance (12.4 S/m^2) and conductance (2.41×10^{-6} F/m^2) values similar to that of the control membrane. Figure 4.4(b) presents the changes in the Nyquist plots during a cleaning process using a NaCl feed solution.

Ho et al. [107] also utilized EIS to monitor the development of biofilm on a RO membrane, specifically the TW30 membrane from Dow FilmTec™. Their findings revealed that the normalized conductance of the diffusion polymerization layer (G_{DP}) exhibited two distinct stages during biofilm formation. The initial stage, spanning from day 1 to day 2 of the experiment, was associated with the accumulation of bacterial cells and the formation of respiration byproducts resulting from bacterial activities. In the subsequent stage, occurring from day 3 to day 5 of the experiment, the focus shifted to the accumulation of extracellular polymeric substances (EPS), which is a primary component contributing to the formation of the biofilm. EIS scans during the filtration of the bacteria (*Pseudomonas aeruginosa*) solution were set at approximately 30 min per scan for up to 5 days. The normalized G_{DP} exhibited a marginal rise from 0.85 to approximately 1.3 on day 1, followed by a rapid drop to 1.0 after 1.5 days, and subsequently decreased to 0.8 after 5 days. On the other hand, the total EPS increased by 13% from day 1 to day 3, and there was a 46% increase in the total EPS from day 3 to day 5. This clearly showed that the EPS produced at day 5 was higher compared to the early stage of biofouling process. The bacteria accumulation on the membrane surface also led to an initial increase in the normalized diffusion polymerization as the bacterial cells and their respiration products proved to be very conductive. During this period, there was a simultaneous occurrence of the initial phase of biofilm development, marked by bacteria adhering to the membrane surface prior to colony formation. As the filtration progressed over a longer duration, once a saturation point was reached, a decline in the G_{DP} was observed. This decline was attributed to the accumulation of a significant amount of EPS. Following

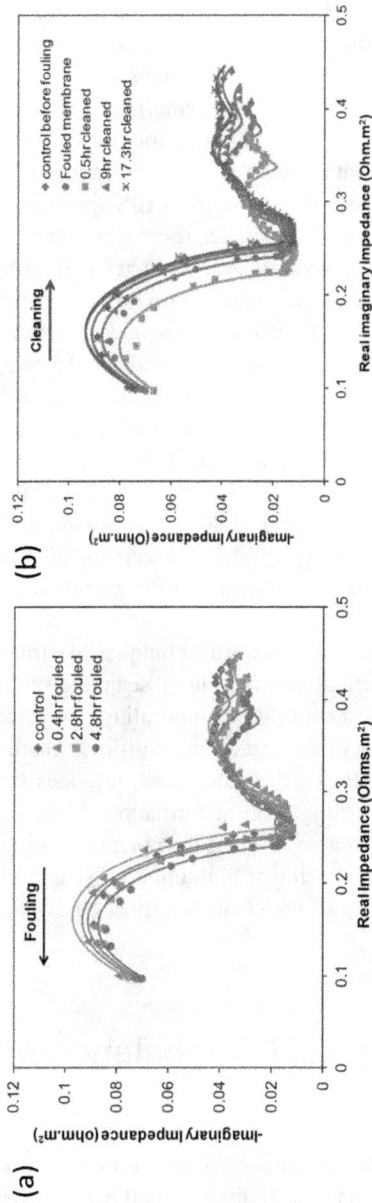

FIGURE 4.4 Nyquist plots representation of EIS data for (a) various times of fouling with silica and alginate and (b) during a cleaning process using a NaCl feed solution [106].

the attachment of bacteria to the membrane surface, their subsequent growth and multiplication played a pivotal role in the development of a biofilm characterized by enhanced structural integrity. It was noted that a tighter EPS matrix resulted in diminished electrical conductivity in the fouling layer, which led to a lower normalized G_{DP} after a period of 1.5 days.

Yeo et al. [109] utilized EIS for a polymeric membrane to assess its performance during FO filtration. This investigation involved both pure water (as feed solution) and draw solution, with KCl concentrations ranging from 0.5 M to 1.5 M. The results showed that the typical EIS spectra of capacitance and conductance as a function of frequencies (between 1 and 10^6 Hz) exhibited a dispersion throughout the spectra over a range of frequencies, indicating the presence of electrically distinct layers with different time constants in the system. Maxwell–Wagner model was then applied to analyze the dispersions observed from the spectra with pure water as feed and 0.5 M KCl as draw solution in the active-layer-facing-draw-solution (ALDS) orientation. Two clearly distinguishable structural components were identified, each characterized by specific capacitance values, i.e., 7.7×10^{-6} F/m^2 and 7.8×10^{-4} F/m^2. These components corresponded to a porous support layer with a thickness of 43 μm and an active layer with a thickness of 80 nm. However, as the concentration of the draw solution increased to 1.5 M, EIS was not able to interpret the interactions between the stationary ion layer and the porous support owing to the diffusion polymerization.

Past studies have often leaned toward techniques like transmembrane pressure (TMP) measurement and confocal laser scanning microscopy (CLSM) to assess a membrane's susceptibility to biofouling. However, these methods offer limited understanding of the underlying biofilm formation processes. The integration of TMP and EIS, on the other hand, provides significant insights into the mechanisms governing biofilm formation. Moreover, in contrast to CLSM, which can be destructive, EIS has the potential for online implementation, enabling real-time monitoring of biofouling in NF and RO systems. It can also facilitate the evaluation of the cleaning efficacy in water and wastewater treatment plants.

4.2.3 Quartz-Crystal Microbalance with Dissipation

Another *in situ* real-time technique used to monitor membrane fouling is based on quartz-crystal microbalance with dissipation (QCM-D). This technique has demonstrated its efficacy in tracking the progression of fouling over time and characterizing the viscoelastic attributes of the fouling layer in the early stage of fouling development. The working principle of QCM-D is based on a microbalance and acoustic wave, measuring changes in frequency and dissipation of a quartz crystal sensor due to the media adsorbed to the sensor [109]. It has exceptional sensitivity (detecting at the ng level) and can detect both the thickness of foulants and their rate of adsorption [110, 111]. When a change of the mass occurs within the thin film, a frequency shift from the fundamental resonant frequency of the crystal (which is corresponded to mass change) is measured accordingly [112]. Under certain conditions when other substances are adsorbed on the surface of the quartz crystal, the amount of the adsorbed mass changes as the natural frequency of the quartz crystal changes. Through the detection of changes in frequency and the dissipation factor, it becomes possible to extract important film characteristics, namely, quality, thickness, viscosity and shear modulus. These parameters are derived using appropriate model fitting techniques.

In order to shed light on the influence of the surface chemical functionality of a TFC membrane on adsorptive fouling, Contreras et al. [110] employed the QCM-D technique. Modifications to a commercial TFC RO membrane were made by incorporating diverse terminating chemical functionalities using self-assembled monolayers (SAMs). The adsorption behaviors of two types of foulants, i.e., BSA and sodium alginate, were investigated. Additionally, they investigated the effects of cleaning on fouling. It was observed that the layers formed by BSA or sodium alginate on various SAMs exhibited comparable normalized dissipation values. This similarity in normalized dissipation indicated that the equilibrium structure of the adsorbed layer was not affected by variations in the membrane surface chemistry. However, alginate layers exhibited much higher measured dissipation change (ΔD)/frequency change (Δf) than BSA, indicating its layers were looser and more elastic. In addition, a two-phase growth was shown by the BSA adsorption, which was corresponded to a notable faster increase in ΔD in the second phase. This suggested changes in molecular conformation or orientation of BSA.

Gutman et al. [113] also investigated the impacts of different model foulants such as bacteria (e.g., *Sphingomonas spp. glycosphingolipids* (GSL), *S. wittichii* and *Pseudomonas aeruginosa*) and lipopolysaccharides (LPS) on the RO membrane surface using QCM-D. Figure 4.5(a) shows two major phenomena with respect to the adhesion of GSL and LPS vesicles. The initial phase occurred when the vesicle was introduced onto the PA surface within the time span of 0 to 120 min. During this phase, a more robust adsorption of GSL vesicles (Δf: −4.5 Hz) was observed in contrast to the adsorption of LPS (Δf: −3 Hz). The subsequent phenomenon which occurred at 120–480 min corresponded to the washing step. This stage involved the detachment or desorption of the vesicles from the PA surface. At the point of maximum frequency, vesicles that were weakly attached to the surface became dislodged. The

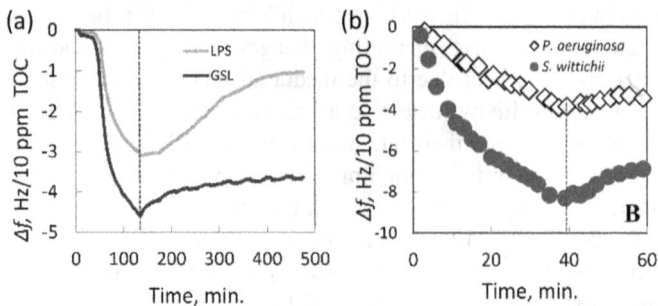

FIGURE 4.5 Decline in frequency of QCM-D sensor upon adhesion of (a) GSL and LPS vesicles with time period of 0–120 and 120–480 min corresponding to adsorption and desorption (washing), respectively and (b) *S. wiitichii* and *P. aeruginosa* EPS (10 mg/L as dissolved organic carbon (DOC)) with time period of 0–40 and 40–60 min corresponding to adsorption and desorption (washing), respectively [113].

QCM-D results showed noticeable differences in the adhesion of LPS versus GSL vesicles, with the latter exhibiting 50% higher adhesion to the PA coated crystals (mimicking an RO membrane surface). This implied not only a greater quantity of GSL adsorbed, but also a strong attachment between the bacteria and the PA surface layer, likely attributed to the hydrophobic interactions.

A comparable pattern was identified for the adhesion tendencies of EPS extracted from *S. wittichii* and *P. aeruginosa*, as illustrated in Figure 4.5(b). Following a 40-min operation, the EPS from *S. wittichii* displayed twice the adhered mass on the sensor compared to *P. aeruginosa* EPS ($\Delta f = -8.3$ Hz/10 ppm DOC for *S. wittichii* versus $\Delta f = -3.8$ Hz/10 ppm DOC for *P. aeruginosa*). Irreversible adhesion was demonstrated, with 89% and 83% of the mass persisting on the sensor after the washing step for *P. aeruginosa* and *S. wittichii* EPS, respectively.

While QCM-D offers the capability to furnish time-dependent insights into molecule-surface interactions governing adsorption, desorption and binding events, its full potential for the characterization of membrane surface fouling remains largely untapped. Particularly in the context of TFC membrane evaluation, there exists a paucity of information. The high sensitivity of the QCM-D technique renders it susceptible to signal attenuation caused by the deposition of additional components or impurities on the sensor. This attenuation could impede accurate interpretation. In addition, the application of the polymeric film (i.e., membrane) onto the sensor through the spin coating technique can present challenges in achieving the optimal coating conditions. Further research is therefore imperative to explore and harness the complete range of capabilities that QCM-D holds for membrane surface fouling analysis.

Conclusions

5

5.1 CONCLUSIONS

The development of PA TFC membranes through the interfacial polymerization technique, pioneered by John Cadotte and his team in the early 1970s, represents a milestone in the field of membrane science and technology. This breakthrough is often considered on par with the historical advancement made by Sidney Loeb and Srinivasa Sourirajan in the 1960s with the introduction of asymmetric membranes made from the phase inversion technique. The interfacial polymerization technique enables the creation of TFC membranes with remarkable characteristics. By employing this method, an ultrathin cross-linked PA layer, often just several hundreds of nm thick, can be formed on top of a microporous substrate. This results in TFC membranes that offer an excellent balance between water flux and salt rejection, especially when compared to the earlier asymmetric cellulose triacetate membranes used in desalination processes.

As of today, the PA TFC membranes have emerged as the dominant choice for industrial applications in NF and RO. The market for NF and RO membranes in water and wastewater treatment is currently experiencing robust growth, reflecting the increasing demand for these technologies in addressing water purification and treatment needs. According to BCC Research, the NF membrane market is expected to exhibit substantial growth, with an anticipated increase from $518 million in 2019 to $1.2 billion by 2024. This projection reflects a compound annual growth rate (CAGR) of 18.2% from 2019 to 2024 (Report no. NAN045C). Additionally, the global market for major components of RO water treatment systems, a crucial application area for TFC membranes, is also on an upward trajectory. It is projected to rise from $11.7 billion in 2020 to $19.1 billion by 2025, with a CAGR of 10.3% during the period from 2020 to 2025 (Report no. MST049G).

DOI: 10.1201/9781032690346-5

The PA TFC membranes also exhibit substantial promise for expanding into osmotically driven processes like forward osmosis and pressure retarded osmosis, and find applications in various other fields such as gas separation, pervaporation and fuel cell. Given these evolving and diverse applications, it would be premature to assert that TFC membrane technology has reached maturity. In fact, numerous opportunities for innovation and advancement still exist, making it an exciting and dynamic area for further research and development.

Developing advanced materials is indeed crucial for enhancing the properties of PA TFC membranes, especially in addressing the flux-selectivity trade-off and ensuring stability under challenging filtration conditions. Notably, advanced characterization techniques play critical roles in further augmenting these membranes. In-depth characterization at the nm scale and even at the molecular level can provide invaluable insights into membrane properties and behavior. These insights contribute to the advancement of water desalination and wastewater treatment technologies. Through advanced characterization techniques, researchers can gain a deeper understanding of factors such as membrane structure/morphology, surface chemistry and interactions with various solutes, which help to judiciously tailor enhancements.

Indeed, the application of advanced characterization methods presents numerous opportunities to tackle the challenges encountered by the PA TFC membranes in water desalination and wastewater treatment. A comprehensive understanding of the underlying physics and chemistry of TFC membrane properties, facilitated by advanced characterization, can significantly expedite the development and adoption of enhanced materials and processes in these critical fields. The benefits of leveraging advanced characterization methods in this context could result in lower energy consumption, improved cost efficiency, increased water flux (without compromising rejection), enhanced fouling resistance, etc.

Overall, the application of advanced characterization methods in membrane research has the potential to drive innovation and lead to the development of next-generation TFC membranes that address current limitations and open up new possibilities in water desalination and wastewater treatment. These advancements are critical in addressing the global challenges associated with water scarcity and pollution while improving the sustainability and efficiency of water treatment processes.

References

[1] J.E. Cadotte, K.E. Cobian, R.H. Forester, R.J. Petersen, Continued Evaluation of in Situ-formed Condensation Polymers for Reverse Osmosis Membranes, Final report, Office of Water Research and Technology. Contract No. 14-30-3298, 1976.

[2] P.W. Morgan, S.L. Kwolek, Interfacial polycondensation. II. Fundamentals of polymer formation at liquid interfaces, *J. Polym. Sci. 40* (1959) 299–327. https://doi.org/10.1002/POL.1959.1204013702

[3] J.E. Cadotte, Interfacially synthesized reverse osmosis membrane, US Patent: 4,277,344, 1981. https://patents.google.com/patent/US4277344A/en

[4] BCC Research, Seawater and brackish water desalination, Market Report: MST052, 2023. https://www.bccresearch.com/market-research/membrane-and-separation-technology/seawater-brackish-water-desalination-report.html

[5] B.-H. Jeong, Y. Yan, A.K. Ghosh, X. Huang, A. Subramani, E.M.V. Hoek, A. Jawor, G. Hurwitz, Interfacial polymerization of thin film nanocomposites: A new concept for reverse osmosis membranes, *J. Memb. Sci. 294* (2007) 1–7. https://doi.org/10.1016/j.memsci.2007.02.025

[6] Y.H. La, R. Sooriyakumaran, D.C. Miller, M. Fujiwara, Y. Terui, K. Yamanaka, B.D. McCloskey, B.D. Freeman, R.D. Allen, Novel thin film composite membrane containing ionizable hydrophobes: PH-dependent reverse osmosis behavior and improved chlorine resistance, *J. Mater. Chem. 20* (2010) 4615–4620. https://doi.org/10.1039/b925270c

[7] S.G. Kim, D.H. Hyeon, J.H. Chun, B.H. Chun, S.H. Kim, Novel thin nanocomposite RO membranes for chlorine resistance, *Desalin. Water Treat. 51* (2013) 6338–6345. https://doi.org/10.1080/19443994.2013.780994

[8] M.N. Putintseva, I.L. Borisov, A.A. Yushkin, R.A. Kirk, P.M. Budd, A.V. Volkov, Effect of casting solution composition on properties of PIM-1/PAN thin film composite membranes, *Key Eng. Mater. 816* (2019) 167–173. https://doi.org/10.4028/www.scientific.net/KEM.816.167

[9] Y. Alqaheem, A.A. Alomair, Microscopy and spectroscopy techniques for characterization of polymeric membranes, *Membranes. 10* (2) (2020) 33. https://doi.org/10.3390/membranes10020033

[10] J. Sun, L.P. Zhu, Z.H. Wang, F. Hu, P. Bin Zhang, B.K. Zhu, Improved chlorine resistance of polyamide thin-film composite membranes with a terpolymer coating, *Sep. Purif. Technol. 157* (2016) 112–119. https://doi.org/10.1016/j.seppur.2015.11.034

[11] M.M.A. Almijbilee, X. Wu, A. Zhou, X. Zheng, X. Cao, W. Li, Polyetheramide organic solvent nanofiltration membrane prepared via an interfacial assembly and polymerization procedure, *Sep. Purif. Technol. 234* (2020) 116033. https://doi.org/10.1016/j.seppur.2019.116033

[12] Madhusha, Difference Between 1H NMR and 13C NMR|Definition, Chemical Shift, Features, Examples and Differences, (2018). https://pediaa.com/difference-between-1h-nmr-and-13c-nmr/ (accessed July 4, 2022).

[13] M. Khajouei, M. Jahanshahi, M. Peyravi, Biofouling mitigation of TFC membrane by in-situ grafting of PANI/Cu couple nanoparticle, *J. Taiwan Inst. Chem. Eng. 85* (2018) 237–247. https://doi.org/10.1016/J.JTICE.2018.01.027

[14] M. Kazemi, M. Jahanshahi, M. Peyravi, 1,2,4-Triaminobenzene-crosslin ked polyamide thin-film membranes for improved flux/antifouling performance, *Mater. Chem. Phys. 255* (2020) 123592. https://doi.org/10.1016/J.MATCHEMPHYS.2020.123592

[15] S.L. Li, P. Wu, J. Wang, Y. Hu, High-performance zwitterionic TFC polyamide nanofiltration membrane based on a novel triamine precursor, *Sep. Purif. Technol. 251* (2020) 117380. https://doi.org/10.1016/J.SEPPUR.2020.117380

[16] A.H. Emwas, R. Roy, R.T. McKay, L. Tenori, E. Saccenti, G.A. Nagana Gowda, D. Raftery, F. Alahmari, L. Jaremko, M. Jaremko, D.S. Wishart, NMR spectroscopy for metabolomics research, *Metabolites. 9* (2019) 123. https://doi.org/10.3390/metabo9070123

[17] H.W. Spiess, 50th anniversary perspective: The importance of NMR spectroscopy to macromolecular science, *Macromolecules. 50* (2017) 1761–1777. https://doi.org/10.1021/acs.macromol.6b02736

[18] X. Kong, V.V. Terskikh, R.L. Khade, L. Yang, A. Rorick, Y. Zhang, P. He, Y. Huang, G. Wu, Solid-state O NMR of paramagnetic coordination compounds, *Angew. Chem. Int. Ed. Engl. 54* (2015) 4753. https://doi.org/10.1002/ANIE.201409888

[19] A.S. Gorzalski, C. Donley, O. Coronell, Elemental composition of membrane foulant layers using EDS, XPS, and RBS, *J. Memb. Sci. 522* (2017) 31–44. https://doi.org/10.1016/J.MEMSCI.2016.08.055

[20] X. Zhang, D.G. Cahill, O. Coronell, B.J. Mariñas, Partitioning of salt ions in FT30 reverse osmosis membranes, *Appl. Phys. Lett. 91* (2007) 181904. https://doi.org/10.1063/1.2802562

[21] O. Coronell, M. ter Horst, C. Donley, Microanalysis of reverse osmosis and nanofiltration membranes, *Encycl. Membr. Sci. Technol.* (2013) 28–31. https://doi.org/10.1002/9781118522318.emst148

[22] Ion Beam Analysis Techniques - RBS Lab, (n.d.). https://sites.google.com/a/lbl.gov/rbs-lab/ion-beam-analysis (accessed August 3, 2022).

[23] L. Lin, C. Feng, R. Lopez, O. Coronell, Identifying facile and accurate methods to measure the thickness of the active layers of thin-film composite membranes - A comparison of seven characterization techniques, *J. Memb. Sci. 498* (2016) 167–179. https://doi.org/10.1016/j.memsci.2015.09.059

[24] L. Valentino, T. Renkens, T. Maugin, J.P. Croué, B.J. Mariñas, Changes in physicochemical and transport properties of a reverse osmosis membrane exposed to chloraminated seawater, *Environ. Sci. Technol. 49* (2015) 2301–2309. https://doi.org/10.1021/es504495j

[25] X. Lu, M. Elimelech, Fabrication of desalination membranes by interfacial polymerization: History, current efforts, and future directions, *Chem. Soc. Rev. 50* (2021) 6290–6307. https://doi.org/10.1039/d0cs00502a

[26] Q. Fu, N. Verma, H. Ma, F.J. Medellin-Rodriguez, R. Li, M. Fukuto, C.M. Stafford, B.S. Hsiao, B.M. Ocko, Molecular structure of aromatic reverse osmosis polyamide barrier layers, *ACS Macro Lett.* 8 (2019) 352–356. https://doi.org/10.1021/acsmacrolett.9b00077

[27] G.M. Geise, H.B. Park, A.C. Sagle, B.D. Freeman, J.E. McGrath, Water permeability and water/salt selectivity tradeoff in polymers for desalination, *J. Memb. Sci.* 369 (2011) 130–138. https://doi.org/10.1016/j.memsci.2010.11.054

[28] A.P. Radlinski, M. Mastalerz, A.L. Hinde, M. Hainbuchner, H. Rauch, M. Baron, J.S. Lin, L. Fan, P. Thiyagarajan, Application of SAXS and SANS in evaluation of porosity, pore size distribution and surface area of coal, *Int. J. Coal Geol.* 59 (2004) 245–271. https://doi.org/10.1016/j.coal.2004.03.002

[29] H.H. Paradies, Particle size distribution and determination of characteristic properties of colloidal bismuth-silica compounds by small-angle X-ray scattering and inelastic light scattering, *Colloids Surfaces A Physicochem. Eng. Asp.* 74 (1993) 57–69. https://doi.org/10.1016/0927-7757(93)80398-X

[30] P.S. Singh, P. Ray, Z. Xie, M. Hoang, Synchrotron SAXS to probe cross-linked network of polyamide "reverse osmosis" and "nanofiltration" membranes, *J. Memb. Sci.* 421–422 (2012) 51–59. https://doi.org/10.1016/j.memsci.2012.06.029

[31] A. Agbabiaka, M. Wiltfong, C. Park, Small angle X-ray scattering technique for the particle size distribution of nonporous nanoparticles, *J. Nanoparticles.* 2013 (2013) 1–11. https://doi.org/10.1155/2013/640436

[32] H. Brumberger, *Modern aspects of small-angle scattering*, Springer, 1995 https://www.amazon.com/Modern-Aspects-Small-Angle-Scattering-Science/dp/904814499X

[33] Y. Liu, H. Wu, H. Zhang, J. Wang, Z. Wang, The evolution of ultrathin polyamide film during molecular layer-by-layer deposition, *Vacuum.* 207 (2023) 111645. https://doi.org/10.1016/j.vacuum.2022.111645

[34] S.A. Sundet, Morphology of the rejecting surface of aromatic polyamide membranes for desalination, *J. Memb. Sci.* 76 (1993) 175–183.

[35] Q. Fu, N. Verma, B.S. Hsiao, F. Medellin-Rodriguez, P.A. Beaucage, C.M. Stafford, B.M. Ocko, X-ray scattering studies of reverse osmosis materials, *Synchrotron Radiat. News.* 33 (2020) 40–45. https://doi.org/10.1080/08940886.2020.1784700

[36] S.E. Bone, H.G. Steinrück, M.F. Toney, Advanced characterization in clean water technologies, *Joule.* 4 (2020) 1637–1659. https://doi.org/10.1016/j.joule.2020.06.020

[37] A.N. Quay, T. Tong, S.M. Hashmi, Y. Zhou, S. Zhao, M. Elimelech, Combined organic fouling and inorganic scaling in reverse osmosis: Role of protein-silica interactions, *Environ. Sci. Technol.* 52 (2018) 9145–9153. https://doi.org/10.1021/acs.est.8b02194

[38] T. Mrozowich, S. McLennan, M. Overduin, T.R. Patel, Structural studies of macromolecules in solution using small angle X-ray scattering, *J. Vis. Exp.* 2018 (2018) 1–8. https://doi.org/10.3791/58538

[39] A.G. Kikhney, D.I. Svergun, A practical guide to small angle X-ray scattering (SAXS) of flexible and intrinsically disordered proteins, *FEBS Lett.* 589 (2015) 2570–2577. https://doi.org/10.1016/j.febslet.2015.08.027

[40] X. Yang, Monitoring the interfacial polymerization of piperazine and trime-soyl chloride with hydrophilic interlayer or macromolecular additive by in situ FT-IR spectroscopy, *Membranes*. *10* (2020) 12. https://doi.org/10.3390/membranes10010012

[41] X. Yang, Controllable interfacial polymerization for nanofiltration membrane performance improvement by the polyphenol interlayer, *ACS Omega*. *4* (2019) 13824–13833. https://doi.org/10.1021/acsomega.9b01446

[42] S.J. Pennycook, P.D. Nellist, *Scanning Transmission Electron Microscopy: Imaging and Analysis*, Springer Science & Business Media, 2011. https://www.springer.com/gp/book/9781441971999 (accessed July 10, 2021).

[43] T.J.A. Slater, A. Janssen, P.H.C. Camargo, M.G. Burke, N.J. Zaluzec, S.J. Haigh, STEM-EDX tomography of bimetallic nanoparticles: A methodological investigation, *Ultramicroscopy*. *162* (2016) 61–73. https://doi.org/10.1016/J.ULTRAMIC.2015.10.007

[44] B.J. Inkson, *Scanning Electron Microscopy (SEM) and Transmission Electron Microscopy (TEM) for Materials Characterization*, Elsevier Ltd, 2016. https://doi.org/10.1016/B978-0-08-100040-3.00002-X

[45] W. Xuan Li, Z. Yang, W. Liang Liu, Z. Hao Huang, H. Zhang, M. Ping Li, X. Hua Ma, C.Y. Tang, Z. Liang Xu, Polyamide reverse osmosis membranes containing 1D nanochannels for enhanced water purification, *J. Memb. Sci. 618* (2021) 118681. https://doi.org/10.1016/j.memsci.2020.118681

[46] R.D. Holbrook, A.A. Galyean, J.M. Gorham, A. Herzing, J. Pettibone, *Overview of Nanomaterial Characterization and Metrology*, Elsevier, 2015. https://doi.org/10.1016/B978-0-08-099948-7.00002-6

[47] S. Sueda, K. Yoshida, N. Tanaka, Quantification of metallic nanoparticle morphology on TiO_2 using HAADF-STEM tomography, *Ultramicroscopy*. *110* (2010) 1120–1127. https://doi.org/10.1016/J.ULTRAMIC.2010.04.003

[48] A.B. Yankovich, B. Berkels, W. Dahmen, P. Binev, P.M. Voyles, High-precision scanning transmission electron microscopy at coarse pixel sampling for reduced electron dose, *Adv. Struct. Chem. Imaging. 1* (2015) 1–5. https://doi.org/10.1186/s40679-015-0003-9

[49] P.J. Kempen, A.S. Thakor, C. Zavaleta, S.S. Gambhir, R. Sinclair, A scanning transmission electron mictroscopy (STEM) approach to analyzing large volumes of tissue to detect nanoparticles, *Micros. and Microanal. 19* (2013) 1290–1297.

[50] R.F. Egerton, Choice of operating voltage for a transmission electron microscope, *Ultramicroscopy*. *145* (2014) 85–93. https://doi.org/10.1016/j.ultramic.2013.10.019

[51] S. Cao, G. Zhang, C. Xiong, S. Long, X. Wang, J. Yang, Preparation and characterization of thin-film-composite reverse-osmosis polyamide membrane with enhanced chlorine resistance by introducing thioether units into polyamide layer, *J. Memb. Sci. 564* (2018) 473–482. https://doi.org/10.1016/J.MEMSCI.2018.07.052

[52] C. Van Goethem, R. Verbeke, M. Pfanmöller, T. Koschine, M. Dickmann, T. Timpel-Lindner, W. Egger, S. Bals, I.F.J. Vankelecom, The role of MOFs in Thin-Film Nanocomposite (TFN) membranes, *J. Memb. Sci. 563* (2018) 938–948. https://doi.org/10.1016/J.MEMSCI.2018.06.040

[53] K.H. Chu, J.S. Mang, J. Lim, S. Hong, M.H. Hwang, Variation of free volume and thickness by high pressure applied on thin film composite reverse osmosis membrane, *Desalination. 520* (2021) 115365. https://doi.org/10.1016/J. DESAL.2021.115365

[54] M. Nystr (3m, M. Lindstr(~m, E. Matthiasson, Streaming potential as a tool in the characterization of ultrafiltration membranes, *Colloids and Surfaces. 36* (1989) 297–312.

[55] V. Freger, Nanoscale heterogeneity of polyamide membranes formed by interfacial polymerization, *Langmuir. 19* (2003) 4791–4797. https://doi.org/10.1021/la020920q

[56] B. Mi, O. Coronell, B.J. Mariñas, F. Watanabe, D.G. Cahill, I. Petrov, Physicochemical characterization of NF/RO membrane active layers by Rutherford backscattering spectrometry, *J. Memb. Sci. 282* (2006) 71–81. https://doi.org/10.1016/j.memsci.2006.05.015

[57] V. Freger, J. Gilron, S. Belfer, TFC polyamide membranes modified by grafting of hydrophilic polymers: An FT-IR/AFM/TEM study, *J. Memb. Sci. 209* (2002) 283–292. https://doi.org/10.1016/S0376-7388(02)00356-3

[58] A.K. Ghosh, B.H. Jeong, X. Huang, E.M.V. Hoek, Impacts of reaction and curing conditions on polyamide composite reverse osmosis membrane properties, *J. Memb. Sci. 311* (2008) 34–45. https://doi.org/10.1016/j.memsci.2007.11.038

[59] V. Freger, Swelling and morphology of the skin layer of polyamide composite membranes: An atomic force microscopy study, *Environ. Sci. Technol. 38* (2004) 3168–3175. https://doi.org/10.1021/es034815u

[60] V.F. Emil Drazevic, Kresimir Kosutic, Permeability and selectivity of reverse osmosis membranes: Correlation to swelling revisited, *Water Res. 49* (2014) 444–452.

[61] S. Karan, Z. Jiang, A.G. Livingston, Sub-10 nm polyamide nanofilms with ultrafast solvent transport for molecular separation, *Science. 348* (2015) 1347–1351. https://doi.org/10.1126/science.aaa5058

[62] R. Xu, G. Xu, J. Wang, J. Chen, F. Yang, J. Kang, M. Xiang, Influence of l-lysine on the permeation and antifouling performance of polyamide thin film composite reverse osmosis membranes, *RSC Adv. 8* (2018) 25236–25247. https://doi.org/10.1039/c8ra02234h

[63] A.L. Carvalho, F. Maugeri, V. Silva, A. Hernández, L. Palacio, P. Pradanos, AFM analysis of the surface of nanoporous membranes: Application to the nanofiltration of potassium clavulanate, *J. Mater. Sci. 46* (2011) 3356–3369. https://doi.org/10.1007/s10853-010-5224-7

[64] W.R. Bowen, T.A. Doneva, J. Austin, G. Stoton, The use of atomic force microscopy to quantify membrane surface electrical properties, colloids surfaces a physicochem. *Eng. Asp. 201* (2002) 73–83. www.elsevier.com/locate/colsurfa

[65] W.R. Bowen, T.A. Doneva, Atomic force microscopy studies of membranes: Effect of surface roughness on double-layer interactions and particle adhesion, *J. Colloid Interface Sci. 229* (2000) 544–549. https://doi.org/10.1006/jcis.2000.6997

[66] H. An, J.W. Smith, B. Ji, S. Cotty, S. Zhou, L. Yao, F.C. Kalutantirige, W. Chen, Z. Ou, X. Su, J. Feng, Q. Chen, Mechanism and performance relevance of nanomorphogenesis in polyamide films revealed by quantitative 3D imaging and machine learning, *Sci. Adv. 8* (2022) 1888. https://www.science.org

[67] C.S. Ong, J.Z. Oor, S.J. Tan, J.W. Chew, Enantiomeric separation of racemic mixtures using chiral-selective and organic-solvent-resistant thin-film composite membranes, *ACS Appl. Mater. Interfaces*. *14* (2022) 10875–10885. https://doi.org/10.1021/acsami.1c25175

[68] A.J. García-Sáez, P. Schwille, Surface analysis of membrane dynamics, *Biochim. Biophys. Acta - Biomembr. 1798* (2010) 766–776. https://doi.org/10.1016/j.bbamem.2009.09.016

[69] P. Nguyen-Tri, P. Ghassemi, P. Carriere, S. Nanda, A.A. Assadi, D.D. Nguyen, Recent applications of advanced atomic force microscopy in polymer science: A review, *Polymers. 12* (2020) 1142. https://doi.org/10.3390/POLYM12051142

[70] P. Eaton, P. Quaresma, C. Soares, C. Neves, M.P. de Almeida, E. Pereira, P. West, A direct comparison of experimental methods to measure dimensions of synthetic nanoparticles, *Ultramicroscopy. 182* (2017) 179–190. https://doi.org/10.1016/J.ULTRAMIC.2017.07.001

[71] K. Tiede, A.B.A. Boxall, S.P. Tear, J. Lewis, H. David, M. Hassellöv, Detection and characterization of engineered nanoparticles in food and the environment, *Food Addit Contam. - Part A Chem. Anal. Control. Expo. Risk Assess. 25* (2008) 795–821. https://doi.org/10.1080/02652030802007553

[72] T. Fujioka, N. Oshima, R. Suzuki, W.E. Price, L.D. Nghiem, Probing the internal structure of reverse osmosis membranes by positron annihilation spectroscopy: Gaining more insight into the transport of water and small solutes, *J. Memb. Sci. 486* (2015) 106–118.

[73] K.L. Tung, K.S. Chang, T.T. Wu, N.J. Lin, K.R. Lee, J.Y. Lai, Recent advances in the characterization of membrane morphology, *Curr. Opin. Chem. Eng. 4* (2014) 121–127. https://doi.org/10.1016/J.COCHE.2014.03.002

[74] V. Pipich, K. Schlenstedt, M. Dickmann, R. Kasher, J. Meier-Haack, C. Hugenschmidt, W. Petry, Y. Oren, D. Schwahn, Morphology and porous structure of standalone aromatic polyamide films as used in RO membranes – An exploration with SANS, PALS, and SEM, *J. Memb. Sci. 573* (2019) 167–176. https://doi.org/10.1016/j.memsci.2018.11.055

[75] Z. Chen, K. Ito, H. Yanagishita, N. Oshima, R. Suzuki, Y. Kobayashi, Correlation study between free-volume holes and molecular separations of composite membranes for reverse osmosis processes by means of variable-energy positron annihilation techniques, *J. Phys. Chem. C. 115* (2011) 18055–18060. https://doi.org/10.1021/jp203888m

[76] Y.H. Huang, W.C. Chao, W.S. Hung, Q.F. An, K.S. Chang, S.H. Huang, K.L. Tung, K.R. Lee, J.Y. Lai, Investigation of fine-structure of polyamide thin-film composite membrane under swelling effect by positron annihilation lifetime spectroscopy and molecular dynamics simulation, *J. Memb. Sci. 417–418* (2012) 201–209. https://doi.org/10.1016/j.memsci.2012.06.036

[77] Y.C. Jean, J.D. Van Horn, W.S. Hung, K.R. Lee, Perspective of positron annihilation spectroscopy in polymers, *Macromolecules. 46* (2013) 7133–7145. https://doi.org/10.1021/ma401309x

[78] J. Li, B. Xiong, C. Yin, X. Zhang, Y. Zhou, Z. Wang, P. Fang, C. He, Free volume characteristics on water permeation and salt rejection of polyamide reverse osmosis membranes investigated by a pulsed slow positron beam, *J. Mater. Sci. 53* (2018) 16132–16145. https://doi.org/10.1007/s10853-018-2740-3

[79] D.M. Davenport, C.L. Ritt, R. Verbeke, M. Dickmann, W. Egger, I.F.J. Vankelecom, M. Elimelech, Thin film composite membrane compaction in high-pressure reverse osmosis, *J. Memb. Sci. 610* (2020) 118268. https://doi.org/10.1016/J.MEMSCI.2020.118268

[80] V. Pipich, M. Dickmann, H. Frielinghaus, R. Kasher, C. Hugenschmidt, W. Petry, Y. Oren, D. Schwahn, Morphology of thin film composite membranes explored by small-angle neutron scattering and positron-annihilation lifetime spectroscopy, *Membranes. 10* (2020) 48. https://doi.org/10.3390/membranes10030048

[81] E.P. Chan, B.R. Frieberg, K. Ito, J. Tarver, M. Tyagi, W. Zhang, E.B. Coughlin, C.M. Stafford, A. Roy, S. Rosenberg, C.L. Soles, Insights into the water transport mechanism in polymeric membranes from neutron scattering, *Macromolecules. 53* (2020) 1443–1450. https://doi.org/10.1021/acs.macromol.9b02195

[82] D. Zheng, D. Hua, A. Yao, Y. Hong, X. Cha, X. Yang, S. Japip, G. Zhan, Fabrication of thin-film composite membranes for organic solvent nanofiltration by mixed monomeric polymerization on ionic liquid/water interfaces, *J. Memb. Sci. 636* (2021) 119551. https://doi.org/10.1016/j.memsci.2021.119551

[83] J.Y. Lee, T.Y. Huang, M.B.M.Y. Ang, S.H. Huang, H.A. Tsai, R.J. Jeng, Effects of monomer rigidity on microstructures and properties of novel polyamide thin-film composite membranes prepared through interfacial polymerization for pervaporation dehydration, *J. Memb. Sci. 657* (2022) 120702. https://doi.org/10.1016/j.memsci.2022.120702

[84] W.S. Hung, M. De Guzman, S.H. Huang, K.R. Lee, Y.C. Jean, J.Y. Lai, Characterizing free volumes and layer structures in asymmetric thin-film polymeric membranes in the wet condition using the variable monoenergy slow positron beam, *Macromolecules. 43* (2010) 6127–6134. https://doi.org/10.1021/ma100559u

[85] J. Lee, C.M. Doherty, A.J. Hill, S.E. Kentish, Water vapor sorption and free volume in the aromatic polyamide layer of reverse osmosis membranes, *J. Memb. Sci. 425–426* (2013) 217–226. https://doi.org/10.1016/j.memsci.2012.08.054

[86] E.P. Chan, A.P. Young, J.H. Lee, J.Y. Chung, C.M. Stafford, Swelling of ultrathin crosslinked polyamide water desalination membranes, *J. Polym. Sci. Part B Polym. Phys. 51* (2013) 385–391. https://doi.org/10.1002/polb.23235

[87] D. Schwahn, H. Feilbach, T. Starc, V. Pipich, R. Kasher, Y. Oren, Design and test of a reverse osmosis pressure cell for in-situ small-angle neutron scattering studies, *Desalination. 405* (2017) 40–50. https://doi.org/10.1016/j.desal.2016.11.026

[88] M. Rezaee, L.C. Tsai, M.I. Haider, A. Yazdi, E. Sanatizadeh, N.P. Salowitz, Quantitative peel test for thin films/layers based on a coupled parametric and statistical study, *Sci. Reports. 9* (2019) 19805. https://doi.org/10.1038/s41598-019-55355-9

[89] M. Nase, M. Rennert, K. Naumenko, V.A. Eremeyev, Identifying traction–separation behavior of self-adhesive polymeric films from in situ digital images under T-peeling, *J. Mech. Phys. Solids. 91* (2016) 40–55. https://doi.org/10.1016/J.JMPS.2016.03.001

[90] F. Luo, J. Wang, Z. Yao, L. Zhang, H. Chen, Polydopamine nanoparticles modified nanofiber supported thin film composite membrane with enhanced adhesion strength for forward osmosis, *J. Memb. Sci. 618* (2021) 118673. https://doi.org/10.1016/J.MEMSCI.2020.118673

[91] Z. Tan, S. Chen, X. Peng, L. Zhang, C. Gao, Polyamide membranes with nanoscale Turing structures for water purification, *Science 360* (6388) (2018) 518–521. https://doi.org/10.1126/SCIENCE.AAR6308/SUPPL_FILE/AAR6308_TAN_SM.PDF

[92] Y. Li, M.M. Kłosowski, C.M. McGilvery, A.E. Porter, A.G. Livingston, J.T. Cabral, Probing flow activity in polyamide layer of reverse osmosis membrane with nanoparticle tracers, *J. Memb. Sci. 534* (2017) 9–17. https://doi.org/10.1016/J.MEMSCI.2017.04.005

[93] F.A. Pacheco, I. Pinnau, M. Reinhard, J.O. Leckie, Characterization of isolated polyamide thin films of RO and NF membranes using novel TEM techniques, *J. Memb. Sci. 358* (2010) 51–59. https://doi.org/10.1016/J.MEMSCI.2010.04.032

[94] X. Liu, W. Li, T.H. Chong, A.G. Fane, Effects of spacer orientations on the cake formation during membrane fouling: Quantitative analysis based on 3D OCT imaging, *Water Res. 110* (2017) 1–14. https://doi.org/10.1016/J.WATRES.2016.12.002

[95] W. Li, X. Liu, Y.N. Wang, T.H. Chong, C.Y. Tang, A.G. Fane, Analyzing the evolution of membrane fouling via a novel method based on 3D optical coherence tomography imaging, *Environ. Sci. Technol. 50* (2016) 6930–6939. https://doi.org/10.1021/ACS.EST.6B00418/SUPPL_FILE/ES6B00418_SI_001.PDF

[96] T.A. Trinh, W. Li, Q. Han, X. Liu, A.G. Fane, J.W. Chew, Analyzing external and internal membrane fouling by oil emulsions via 3D optical coherence tomography, *J. Memb. Sci. 548* (2018) 632–640. https://doi.org/10.1016/J.MEMSCI.2017.10.043

[97] S. Park, T. Nam, J. Park, S. Kim, Y. Ahn, S. Lee, Y.M. Kim, W. Jung, K.H. Cho, Investigating the influence of organic matter composition on biofilm volumes in reverse osmosis using optical coherence tomography, *Desalination. 419* (2017) 125–132. https://doi.org/10.1016/J.DESAL.2017.06.002

[98] L. Fortunato, S. Bucs, R.V. Linares, C. Cali, J.S. Vrouwenvelder, T.O. Leiknes, Spatially-resolved in-situ quantification of biofouling using optical coherence tomography (OCT) and 3D image analysis in a spacer filled channel, *J. Memb. Sci. 524* (2017) 673–681. https://doi.org/10.1016/J.MEMSCI.2016.11.052

[99] L. Fortunato, T.O. Leiknes, In-situ biofouling assessment in spacer filled channels using optical coherence tomography (OCT): 3D biofilm thickness mapping, *Bioresour. Technol. 229* (2017) 231–235. https://doi.org/10.1016/J.BIORTECH.2017.01.021

[100] T.A. Trinh, W. Li, J.W. Chew, Internal fouling during microfiltration with foulants of different surface charges, *J. Memb. Sci. 602* (2020) 117983. https://doi.org/10.1016/j.memsci.2020.117983

[101] Y. Gao, S. Haavisto, W. Li, C.Y. Tang, J. Salmela, A.G. Fane, Novel approach to characterizing the growth of a fouling layer during membrane filtration via optical coherence tomography, *Environ. Sci. Technol. 48* (2014) 14273–14281. https://doi.org/10.1021/ES503326Y/SUPPL_FILE/ES503326Y_SI_001.PDF

[102] M.U. Farid, J. Guo, A.K. An, Bacterial inactivation and in situ monitoring of biofilm development on graphene oxide membrane using optical coherence tomography, *J. Memb. Sci. 564* (2018) 22–34. https://doi.org/10.1016/J.MEMSCI.2018.06.061

[103] J.M. Kavanagh, S. Hussain, T.C. Chilcott, H.G.L. Coster, Fouling of reverse osmosis membranes using electrical impedance spectroscopy: Measurements and simulations, *Desalination.* *236* (2009) 187–193. https://doi.org/10.1016/J. DESAL.2007.10.066

[104] L.N. Sim, Z.J. Wang, J. Gu, H.G.L. Coster, A.G. Fane, Detection of reverse osmosis membrane fouling with silica, bovine serum albumin and their mixture using in-situ electrical impedance spectroscopy, *J. Memb. Sci.* *443* (2013) 45–53. https://doi.org/10.1016/J.MEMSCI.2013.04.047

[105] J.S. Ho, L.N. Sim, R.D. Webster, B. Viswanath, H.G.L. Coster, A.G. Fane, Monitoring fouling behavior of reverse osmosis membranes using electrical impedance spectroscopy: A field trial study, *Desalination.* *407* (2017) 75–84. https://doi.org/10.1016/J.DESAL.2016.12.012

[106] L.N. Sim, J. Gu, H.G.L. Coster, A.G. Fane, Quantitative determination of the electrical properties of RO membranes during fouling and cleaning processes using electrical impedance spectroscopy, *Desalination.* *379* (2016) 126–136. https://doi.org/10.1016/J.DESAL.2015.11.006

[107] J.S. Ho, J.H. Low, L.N. Sim, R.D. Webster, S.A. Rice, A.G. Fane, H.G.L. Coster, In-situ monitoring of biofouling on reverse osmosis membranes: Detection and mechanistic study using electrical impedance spectroscopy, *J. Memb. Sci.* *518* (2016) 229–242. https://doi.org/10.1016/J.MEMSCI.2016.06.043

[108] N. Zhang, M.A. Halali, C.F. de Lannoy, Detection of fouling on electrically conductive membranes by electrical impedance spectroscopy, *Sep. Purif. Technol.* *242* (2020) 116823. https://doi.org/10.1016/J.SEPPUR.2020.116823

[109] S.Y. Yeo, Y. Wang, T. Chilcott, A. Antony, H. Coster, G. Leslie, Characterising nanostructure functionality of a cellulose triacetate forward osmosis membrane using electrical impedance spectroscopy, *J. Memb. Sci.* *467* (2014) 292–302. https://doi.org/10.1016/J.MEMSCI.2014.05.035

[110] A.E. Contreras, Z. Steiner, J. Miao, R. Kasher, Q. Li, Studying the role of common membrane surface functionalities on adsorption and cleaning of organic foulants using QCM-D, *Environ. Sci. Technol.* 45 (2011) 6309–6315. https://doi. org/10.1021/ES200570T/SUPPL_FILE/ES200570T_SI_001.PDF

[111] X. An, K. Zhang, Z. Wang, Q.V. Ly, Y. Hu, C. Liu, Improving the water permeability and antifouling property of the nanofiltration membrane grafted with hyperbranched polyglycerol, *J. Memb. Sci.* *612* (2020) 118417. https://doi. org/10.1016/J.MEMSCI.2020.118417

[112] G. Rudolph, A. Hermansson, A.S. Jönsson, F. Lipnizki, In situ real-time investigations on adsorptive membrane fouling by thermomechanical pulping process water with quartz crystal microbalance with dissipation monitoring (QCM-D), *Sep. Purif. Technol.* *254* (2021) 117578. https://doi.org/10.1016/J. SEPPUR.2020.117578

[113] J. Gutman, M. Herzberg, S.L. Walker, Biofouling of reverse osmosis membranes: Positively contributing factors of Sphingomonas, *Environ. Sci. Technol.* 48 (2014) 13941–13950. https://doi.org/10.1021/ES503680S/SUPPL_FILE/ES503680S_ SI_001.PDF

Index

Pages in *italics* refer to figures and pages in **bold** refer to tables.

For Product Safety Concerns and Information please contact our EU
representative GPSR@taylorandfrancis.com
Taylor & Francis Verlag GmbH, Kaufingerstraße 24, 80331 München, Germany